*Performing Arts International*
1996, Vol. 1, Part 1
Reprints available directly from the publisher
Photocopying permitted by license only



# Contents

*Performing Arts International*
1996, Vol. 1, Part 1, p. 1
Reprints available directly from the publisher
Photocopying permitted by license only



# Introduction

Michael Bridger

In this first issue of *Performing Arts International* our aim has been to give an impression of the range of issues and types of writing that the journal will cover in succeeding issues. Dance, drama and music all have a presence here, though in different ways and contexts, and to differing extents. Some articles are overtly transdisciplinary, others more specialised in their focus. Taking another perspective, some articles are at a fairly abstract level of theory, while others are more in the nature of practitioners' reflective notes. The international dimension is present too, stemming in part from initiatives and contacts related to Middlesex University's interest in Asian developments in, and attitudes to, the performing arts. Styles of writing vary too, and a conscious decision has been made not to edit out all the characteristic flavour of contributors writing in what for them is a foreign language. It is not easy to characterise a miscellany such as this with a single word, but 'Hybridity' seems not too far off.

While future issues will in general be less eclectic in their agendas, there is more lying behind this issue than just to present a trailer of future directions. Never before has the world of performance been more richly diverse, and more readily accessible on a global scale. Whatever its intellectual and communicative aspirations, the launch of this journal is timely as a celebration of that diversity.

*Performing Arts International*
1996, Vol. 1, Part 1, pp. 3–20
Reprints available directly from the publisher
Photocopying permitted by license only



# Collaboration: Process or Product, Means or End, Democracy or Demagogy?

David Hughes

An exploration of "the nature of collaboration in its creative, practical, aesthetic and political aspects through both the language in which it is thought and idealised, and the languages in which it is conducted".

KEY WORDS: Collaboration, hybridity, choreography, composition, post-modernism, Cage/Cunningham/Rauschenberg, synthesis, transformation.

## *The Exchange Reports, Seminar Papers and the Current Article*

The current article is mainly based on a report which I wrote as documentist of the Composers and Choreographers Exchange 1993, held at London's South Bank Centre. That report incorporated the text of a theoretical paper. A version of that paper was then delivered under the title 'Is Collaboration a Dirty Word?' as part of the Rhythm Method Seminar, also at the South Bank Centre, in February 1994. The 'Dirty Word' paper became the text of the introductory paper to the Exchange in November 1994. A version of that appeared in 1994's Report. The paper published here shows all the traces of those previous versions and has some new thinking which is the product of the current rewriting and also conversations with Robert Wilson and Hans Peter Kuhn during the opening of their installation, *H.G.* at the Clink Street Vaults, London, on September 11th, 1995.

## *The Exchange Itself*

The aim of the Exchanges which I documented in '93 and '94 was to analyse the nature of collaboration in its creative, practical, aesthetic and political aspects through both the language in which it is thought and idealised and the language in which it is conducted.

The power relations in dance/music collaborations usually favour the choreographer, it was felt. The '93 Exchange wanted to examine the composer's role in that power structure and to see what could be learned about self-generated performance from the ways that dancers work.

The Exchange is now an annual event at which 10 composers and 10 choreographers meet for around one week and engage in a variety of seminars, workshops, classes and work sessions. The project of the Exchange has come to be an ongoing analysis of the nature of 'collaboration'.

## *Starting Points for the '93 Exchange*

My work began from a series of conversations with the Exchange Director, the composer, Glyn Perrin. Collaboration, went our initial premise, is a word which can mask traditional power relations giving the illusory gloss of new and radical forms of working. It was the role of the Exchange to help the participants understand how collaborations work and whether their expectations of them were realistic contractually, politically, creatively, aesthetically and interactively.

It was hoped that analysing the nature of collaboration would help change for the composers the models for initiating, funding, making and showing work. Amongst the reasons why the models need to change is to redress the imbalance in the current funding and creative initiatives. Whilst choreographers are able to initiate collaborations with composers, it is extremely difficult within current funding structures – and within the traditional, Western orchestral system – for composers to initiate collaborations with choreographers. There was a sense that otherwise the power would always lie with the traditional initiator – the choreographer.

## *I. The Language of 'Collaboration'*

What does the metaphoric language that surrounds collaboration tell us? We talk of marrying bits of performance to music, for example, but also of marrying together collaborators. Marriage suggests both a model of partnership and exploitation, the expression of desire and the product of a formal arrangement. But despite the very real negative aspects of this metaphor, potential collaborators – latent collaborators? – still regard it through rosy spectacles, as though marriage did in fact represent the height of romantic aspiration. There was a wilful blindness, then, that we perceived and wished to counter-balance.

'Collaboration' covers conditions that stretch all the way from close blow by blow communications, each advance in one artist's thinking being reflected in the other's, to the hands off kind of collaboration instituted by Cunningham where one conversation regarding the duration of the projected price might have sufficed.

'Collaboration' is like the concept 'game', a wide family term where there is infinite variety, no core, ideal form but which covers many small groups of games which have resemblances to other related groups. 'Collaboration' covers different kinds of ways of working together from the closest and most generous to the most ragged arsed and acrimonious combat to get one's ideas into the show.

When we speak of collaboration we sometimes talk of families, of kinship, of tribes, of affairs and liaisons. The liaisons of collaboration seem to be perpetually dangerous, the matings incestuous. Are there taboo structures that collaboration violates? There are, of course, the traditional, hallowed totems of ownership and property, authorship and racial purity of forms which collaboration transgresses.

There was talk early in the Exchange of bolting together disparate aesthetic elements, of making, of devising. In the past we might well have talked about 'group devised pieces' where now we talk of 'collaboration'. Does the new term mark any substantive difference in the process or the product, we wondered?

It has been suggested that the term marks a very real change in the way that choreographers make work, choreographers drawing much more heavily on the contributions of their dancers and perhaps, much more concerned with visual and aural elements for which they need to draw on other kinds of expertise. The star system in classical ballet and the orchestral system now means that dancers and musicians can have more say in the making of work, but it has only been institutionalised in music since the 60s when Henze, Stockhausen and others made works where the players were invited to make choices and contributions. It is much more recent a phenomenon in ballet. Historically, is it marking the increased importance of the directorless performance company, the ensemble which is musician-led (such as the Kronos)?

In dance we have seen an interest in new aesthetics, in new languages, in a movement towards other forms and away from the strictly terpsichoric. There has been a move away from the decorative mode in dance which necessitated or presupposed the unison and uniform group, towards an interest in the individual movement profiles and characters of individual dancers. The search for new dance languages has gone on in some respects in the invitation to dancers to improvise and develop their own material which is then incorporated into the whole. This has frequently manifested in a more theatrical kind of dance, a dance theatre, and the art-form models, roles and job-descriptions of cinema (editor, director) and art (collagist) become as useful in describing the processes and products as the technical language of choreography.

The situation of collaborators making material that doesn't fit each others aesthetic preferences happens all the time. Why would people want to do it if their aesthetics and ideas are going to be systematically affronted and challenged? Perhaps because they have a belief in new forms of art which are not single or hierarchical, but hybrid, fissured, many-faceted. Perhaps because they perceive some skill or style in a collaborator that they wish to integrate into their own work.

Collaboration has become a fashionable vogue and a voguish fashion, and is in a sense imposed on artists by funders who, importantly, have recognised that people work in this way and have established funds to support it. However, it now means that people look for ways of getting that money and so look for collaborators for funding purposes rather than because they necessarily really want to work with them. It has become so fashionable that a set designer may now be called a visual collaborator. But why the move to calling design and composition functionaries collaborators when the art form and power relations are unchanged?

An important misunderstanding, evident in funders' initiatives, is that the formal aspects of art can be renewed, revitalised and radicalised through hybridisation. To effect this hybridisation, this making of new artforms from combinations of existing ones (the notion of 'combined' arts has been influential in institutional thinking about performance and live art), funders set the criteria of certain funds to include a promiscuous round of couplings between artists – new collaborations. I pick up this issue – that collaboration results in hybrid or combined art – in section III of this article.

The New Collaborations Fund, administered by the Arts Council's Combined Arts Unit, whilst a valuable addition to funding strategies, insists that collaborations are new. If they become established the applicant is transformed into a company for the purposes of assessment. This means that artists have constantly to look for new collaborators to satisfy these requirements. The historic models show us that when you find the right collaborator the partnership goes on for years, for a lifetime, perhaps. This shows a paradox in the understanding of what collaboration can be and fosters the need for novelty rather than mature understanding.

Collaboration can frequently look like a marketing gimmick. There is a fashion for wild matings of artists and forms. And yet there is a paradox, because culturally we celebrate still the individual, the signature that unifies, the author. There is little or no talk of collaborators on a novel, a poem, relatively little of collaborators on film, on the composition of classical music. The output unified by an organising sensibility is still our cultural ideal. It is easier to work back to a unifying genius or sensibility in our critical activity, that reassures our humanist desire for 'man-centredness'. But the surrealists initiated the idea of the text from automatic writing, the collective text and so on, and of course the celebration of the distinct author is not a natural given, it is historically specific. There was a time when authorship was not accredited in this way, when there was no distinction between scribes and originators of texts. Until the 19th-century periodical and newspaper articles were not given by-lines. They were anonymous. Culture makes individual authors into unique signatures. An artist needs to be recognised to be able to continue within funding and promotional structures.

The whole music sub-culture of live mixing in the dance gig takes away the notion of the single author, things are re-authored in the mix, but now we have a new kind of artist, the DJ who mixes, and they are becoming highly paid unifiers with a marketable brand name and brand identity. However far we move from the unifying sensibility model in practice, it seems it is always reintroduced in some way.

There is a further interesting paradox in the situation where a product has to be distiguished by its individuality (like the signatures written on the bodywork of new cars) and vogue for collaboration which means that in sense any element could be substituted by another and the work has a potentially shifting identity.

## *II. Defining 'Collaboration' I*

The Oxford English Dictionary defines 'collaborate' in the following ways:

1: Work jointly at literary or artistic production.
2: Co-operate traitorously with the enemy.

The Oxford Paperback Dictionary construes it as 'to work in partnership', whilst the Collins Softback Dictionary tells us that 'collaborate' is often followed by 'on' or 'with'.

Immediately we can see the disparity between the notion of an 'ideal', 'true', 'perfect' collaboration (partnership[1]) which suggests an equality, a non-hierarchical structure through the term 'jointly', and the realities of working with other people at artistic production (betrayal, treachery), the product and the process, both implied by 'production'. A definition, then, which says nothing about pretexts for work (aesthetics, source material, politics, mutual respect), nor about the role of the initiator (who pays whom, the division of labour, public credits and so on).

'Collaboration' also points to a situation of coercion, working under duress, of compromise and self-deprecation (or perhaps self-aggrandisement), in a power structure. A betrayal of principles and forms in a hierarchy of personalities and art forms that relegate an artist or an art form to a sub-dominate position in a formal or professional structure. Which is to say that one art form or artist may dominant author(is)ing position.

What we have then is one word which covers the ideal equality of forms and artists, and the opposite, a reinforcing of the traditional hierarchies of forms and functionary. What is clear is that the term *can* mask exploitative situations just as easily as marking situations of equality – can mark both democracy and demagogy. The challenge to the ideal of collaboration is frequently one of personality, but it is equally that we still hold culturally to the notion that there are such hallowed traditions as authorship and dominance of forms (usually everything is subjected to the tyranny of the text, the human body, presences rather than absences).

If traditionally within theatrical and ballet performance we can point to a demagogy of workers and forms within a hierarchy, we can also, in the light of work done in dance, performance art and the hybrid arts in general, point to a number of other models.

## *Models*

There are many models of collaboration. The self-evident ones include working in a social group developing the work under hand, working in isolation and bringing the different works together at the end, dialectical progression, parallel progression, themed and unthemed, but there are many other models: Tadeusz Kantor talked of his collaboration with fellow-Pole, the surrealist playwright, Witkiewicz, although Witkiewicz was long dead. It is that the texts of Witkiewicz which Kantor uses as the verbal and vocal ingredient in his stage compositions functions as autonomous entities, that he partly means. The stage work is not an interpretation of the text. Kantor designed the whole thing, but each signifying track – sound, costume, movement, etc. – did not support an interpretation of a text, they said different things. But those clashing tracks cohere under an unifying aesthetic umbrella – Kantor's expressionistic, visual style signature.

The thrust of Kantor's assertion about collaborating with Witkiewicz is that the author as unique originator of the text is a cultural fiction: all texts reflect other texts, other modes of writing, styles which are recognisable and bring credibility to a work. The author is steeped in a history, a tradition and that is manifest in their writing. Kantor was constructed by Witkiewicz in important ways: Witkiewicz was an influential force on 20th-century Polish art across all the art forms. All previous texts enter into the production of subsequent texts, they are all collaborators. Kantor indicates in his assertion a foregrounded and special relationship with texts of Witkiewicz: that there is a sustained interrogation of Witkiewicz's texts, that the stage work effects a kind of *dialogue* with the texts of Witkiewicz, but that they are considered as whole things in their own right that sit alongside the visual 'art' made by Kantor. It is a mark of the equal status of the art and the artists.

There were John Cage, Merce Cunningham and Robert Rauschenberg who worked, apparently with relative freedom but under an aesthetic umbrella, Cunningham's signature. Each artform and each artists apparently, though not actually, autonomous.

Rock groups typically function in a very collaborative way, but within the form, each musician making their own kind of gesture which combines to create the group identity. It is why group members need to make solo work. It is difficult to control the overall sound in this situation. Frank Zappa made a point of stressing to his musicians that they were there to play his music his way. It needed to be said because the conventions imply a great deal of creative freedom for each member and Zappa was also known at auditions to encourage the musician to play their most outrageous and personal styles.

A visit to the Exchange '93 by the production team of *Therese Raquin* (then playing at London's Young Vic Theatre), threw up some interesting observations. Director Julia Bardsley, composer Andrew Poppy and set designer Aldona Cunningham unwittingly highlighted a situation where the power hierarchy of theatre was still in place, but the functionaries were now referred to as collaborators. It was never clear quite why: trivially it appeared to mark the fact that the production team had long conversations where they developed the rational and symbolism of the scenic, action and audio elements, but here was agreement that finally it was Bardsley (the director's) show. Part of Bardsley's media profile is that she brings to theatre lessons learned from performance art and visual theatre. So the terminology hints at those antecedents. But this was the clearest moment of realising that there had been a significant change in terminology. This infers a change in expectations, assumptions, working relations and forms, but in practice there may be no difference between the 'collaborative' situation and the typical theatre hierarchy. One question is how much the increased influence of performance art models has to do with the change in vocabulary? The increased co-operation of artists as opposed to a hierarchy where a director merely uses the services of functionaries with particular skills. The perception of the marketing cache attached to collaboration may also be having its effect.

The term collaboration in light of the Bardsley Case Study seems to function like any other ideological term. It masks the same power and production relations that always existed in theatre. It also masks the antagonism of interests that typically inhere.

## *Defining 'Collaboration' II*

We proposed for ourselves a definition of collaborative process which we could test out during the Exchange, because the term is used to cover so many things it's difficult to get a grip on it. This isn't the dictionary definition any longer, but the popular, commonly assumed one, the one that sets up all the unrealistic expectations: 'collaboration is the equal freedom of a group of artists to create and have influence in the decision-making process in the creation of an original art work'. This definition embraces crucially the notion of an equal partnership (which is the kind of language used in funding application forms), the democratic right of all involved to equal part in the decision making process. It also excludes working on a pre-existing text as interpreters of that text. What sometimes gives 'collaboration' a bad name is that the power relations can be manipulated by the unscrupulous demagogue so that an overriding idea can function as a tyrannical text.

Immediately the definition of collaboration I've just given will seem impossibly ideal. If it works at all then there will be degrees of these conditions in each grouping we could call collaborative. It's what happens when we reach the extreme where there is very little of this definition operating, but yet the process is still called collaboration, that's worrying. I doubt if the fullest extreme the definition suggested is ever achieved or is possible.

This understanding of collaboration gives rise to unrealistic expectations in both initiators and initiates. Some of the difficulties are that initiators may feel their collaborators will have the same degree of commitment to the project as they do, whilst the initiates might in fact only see themselves as employees. Alternatively, initiates may feel that the term collaborator endows them with equal rights of decision making, freedom to create, payment, billing and so on, and yet they may sadly discover that in fact they have only been employed as functionaries, to produce something to order.

What is equality in this context? Equal talent, skill, experience, amount of the show which is an artist's product? Hans Peter Kuhn suggests only 50% of your own ideas find their way into a collaboration![2]

We know that time makes it unlikely that all parts of the process and every decision will be open for debate. We know that collaborations are frequently set up by key people in a project and that the work is essentially their work, but that the term collaboration marks a recognition of the high status and artistic quality of a partner. It may be that the previous work of an artist is considered to be just the right kind of thing to fit into one's conception of a new project and the invitation to collaborate is an invitation of them to continue making their kind of work in this new situation. This leads to that familiar conversation overheard in rehearsal when the music finally arrives, late of course, and you say why isn't it like your last piece and the composer says, sorry I've moved on from that, not interested in miking up the stage floor any more, I'm into twelve tone and glass harmonicas now.

In collaborations it is usually the case that one artist sets the terms in which other artists work, practically, aesthetically, contractually (and, importantly, over what length of time) and gives them their degrees of freedom, implicitly claiming total freedom for him/herself. And that key employing artist will probably be writing the contracts that define the business and financial grounds. The question of how the collaborators are billed on the poster and programme may not be considered at all at this stage, or it might be accepted that the company name or the employing artist's name will take the higher profile. Rightful authorship (or co-ownership) of a piece is not necessarily equally acknowledged, the final production may belong to a company other than the group of people who made it, there may be inequalities of payment for equal labour or equalities of payment for unequal labour. The amount of time that each artist takes to create might be different and the amount and kind of labour necessary, also different. Where someone produces their work will place them closer to the seat of decision making or further away. And someone might have final rights of veto in the editing suite or during the technical rehearsal, which was where Cunningham famously changed some Rauschenberg designed constumes deemed inappropriate.[3]

Whilst many individual signatures may appears on elements in the collage, the overall structure and defining characteristics, Cunningham's choreography, Wilson's visual aesthetic, many well be the overriding and dominant signature. In marketing the work, the name of the most saleable collaborator may take precedence.

We are familiar now with production companies and curators forging collaborations between artists which they employ – this is called commissioning, I believe – to create a unique piece of product for them to market on the novelty ticket. Usually as a site-specific event to add spice and the adventure of travel to some draughty, isolated wharf. The recent disappointing installation by Robert Wilson and Hans Peter Kuhn at the Clink Street Vaults may well be one such confection.

Collaboration, then is a term which may be used as a marketing ploy to indicate the novelty of a production team. This marketing gimmick, and the sense that funding bodies want new collaborations, can militate against the maturing of working groups of artists who come to know each other and expand the freedoms they take and the influence they share. A knowledge that can make the process easier or more fruitful. Hans Peter Kuhn, talking about his work with Robert Wilson, relates how they fought madly to begin with because Kuhn, used to working with Peter Stein, had got used to thinking about what the director wanted, not what he as a sound artist wanted.[4] Eventually, he discovered that the solution to working with Wilson, the solution to finding what it was that Wilson wanted, was simply to do what he, Kuhn, wanted to do himself. They have now worked together on quite a few projects perfectly happily, apparently. Such a process takes time.

It is suspect when funding schemes force an artist to look for collaborators from particular sectors of the art and cultural communities. Particularly as this suggests that it is the funders and promoters who are defining the product, setting the agendas, not the artists. Encouraging new groupings can be seen as a way of shaking people out of stale formulae, but, imagine if Cage and Cunningham had only been allowed one collaboration? What counts is the inventiveness and integrity of the artists.

The word collaboration covers such a wide range of phenomena as to be almost useless, but it marks a difference in vocabulary and in political thinking from terms such as 'co-operative', 'collective' or 'ensemble'. The naming of 'artists' within collaborations with other 'artists' is consistent with the tendency throughout the '80s to give all jobs inflated and inflating new names. In one way it aims to even out the social contours. Everyone is important, everyone may have aspirations. Collaboration figures prominently in the language of the EC across all fields of activity from banking of culture, as well as agriculture. No-one is disadvantaged in the new collaborative environment. It is a classless society of equal partnerships.

What we have seen over the last ten years is an increased concern for the whole work, the integrated work, the work that pays as much attention to its spatial, visual and audio elements as to text and live performance. We do need to recognise the attempt to create the integrated hybrid work. If the term collaboration marks this respect for the human contributions, the necessary infrastructures, the attempt to balance and work with different art forms within performance, then collaboration may be used strategically and politically.

## *Collaboration Under Erasure*

It will come as no surprise, then, to read that in this paper I use the term 'collaboration' under erasure, by which I mean that whilst we have the work and sort of know how to use it, it is not ideal. It covers many different kinds of situation and when we use it we know that it is highly qualified, acknowledging that it can mistakenly be taken to mean a zone of blissful synergy, where a number of fully realised wholes becomes parts of a greater whole. No pain, no compromise – everything was beautiful. We know how to use it but we're not really sure that the situations we use it in are really collaborations. It covers all kinds of situations where artists work together. The main idea is that there is a marriage of equals, equal artists and equal art forms. But things change when people and art forms are brought into the same field, the zone of their overlapping and combative fields of force. What are the changes that happens to artists and art forms when they enter the same field, the same space?

The Exchange is a microcosm and microscope to analyse that question.

## *III. Confusions: Collaborative Works – Hybridity and Collaboration*

There are a number of reasons why the word 'collaboration' might have become besmirched. One has to do with the processes of working together, one has to do with a confusion between the process and the product. The sense that collaborative process will result in hybridity of form is implicit in the much heard formulation 'collaborative work'. There is frequently a confusion, then, between the notions of collaboration and hybridity.

We have to distinguish between the process of collaboration on the one hand which is a human and social situation, and hybrid works on the other which is a formal condition. I don't think there is any such thing as a collaborative work, that's a category mistake. I propose that works generated by collaborative processes do not necessarily emerge as hybrid art forms, and hybrid works, when they exist, have not necessarily emerge as hybrid art forms, and hybrid works, when they exist, have not necessarily been the products of collaborative processes. We can test this hypothesis by asking if various kinds of collaborative work necessarily have to follow from a collaborative process.

Against the traditional dominant/support model when all the elements simply combine to reinforce the interpertation of a text, for example, I want to suggest three kinds of hybridity in art: juxtaposition, synthesis and transformation.[5]

### *Juxtaposition*

I can characterise Levinson's position on juxtaposition by the following: juxtapositional works see the products or objects of two or more art forms shown together in the same field. The objects, forms, events are still recognisable as themselves, as individual and complete examples of products of art forms – painting, a dance, a sculpture, a film. The elements add up to the total event but could be separated and shown individually.

In his lecture[6] Robert Wilson refers to his apartment block metaphor and structure for accommodating and juxtaposing the work of collaborators. He creates the overall structure or architecture, a kind of 'apartment block' in which his collaborators make their own homes. However, Hans Peter Kuhn suggests that this placing of work alongside other work is not 'true' collaboration and that some synthesis of ideas is necessary for the activity to qualify as collaboration. It is clear that Kuhn believes that the nature of his own collaboration with Wilson is of the synthetic kind, two sets of ideas that influence each other.

In the work of Cage, Cunningham and Rauschenberg, this state of affairs was taken up and used strategically. Cage made sound quite independently, whilst Rauschenberg made costumes and settings. Cunningham made the movement according to its own dynamics and imperatives and timings. Perhaps it was fortunate that Cage's music was not particularly pulse-based or there may have been practical and aesthetic clashes between music and dancers/dance.

It was certainly not the case, however, that each artist in the Cage/Cunningham/Rauschenberg collaboration had total freedom to affect the space of the other. The effect of one field on the other was regulated. The notorious Rauschenberg costume was changed at the last moment for two reasons, they were too heavy for the dancers to effect their leaps and their tone didn't fit with the emerging tone of the piece. The elements may be juxtaposed, they may in a Cunningham piece be visibly and audibly distinguishable, but they fit together in a collage that is authorised by Cunningham. And the term author is there not by accident. Juxtaposition is Cunningham's process of authoring. And whilst authoring does not here suggest responsibility for creating each element, it does for making it into a whole – the rhythmic congruence of disparate elements.

A work such as Cage's *Roaratorio, An Irish Circus for Finnegans Wake* is similarly authorised and authored by Cage and has an unmistakable signature. The all-over focus and the sound of his voice are major elements in that signature, yet that also is made up of the improvisations of disparate parties of singers, musicians, etc., and the juxtapositions of prepared dance sequences. Both Cage and Cunningham accept that each viewing and listening point is going to be relative (spatially and culturally) and that there will be different stories created at each point of reception of the work. The audience then becomes a kind of collaborator, making the meanings of the work, re-authoring the work as text, through site-specific context. It is not closed and presented for consumption, it remains relatively open so that the plays of signification across the fields of sound, movement and architecture (the division of the space by settings, lights and so on) are constituted by the observer. These pieces resist one interpretation. Wilson at once seems to take on board this re-contextualising, reframing effect of juxtaposition, whilst also attempting to delimit that interpretative freedom by constructing the visual and spatial limits of the work.

Robert Wilson and Philip Glass's *Einstein on the Beach* is quite clearly made up of separate art forms: there is sculpture (an object will simply sit on the stage floor, as it might in a gallery, for a long period of the show), bodies will perform self-contained movement quite divorced from any other element, there is song, there is music. The whole thing operates by juxtaposing these elements and inviting us to view things separately and in combinations. The music has a certain autonomy, but within his overall scheme. Philip Glass's sound signature becomes something we recognise as part of Wilson's signature, it is absorbed into it, but aesthetically it remains coherent within itself. It is releasable as an autonomous recording. The work as a whole employs the music as a flux that gives a coherence of time, a river that carries the event forward. Wilson's work contains a number of the hybrids I refer to here, it is a hybrid of hybrids. The network of connections which develops between the images[7] militates against reductivist readings.

We saw in Sue MacLennan's[8] workshop in the '93 Exchange how juxtaposition can operate as a structuring strategy within a form, setting pieces of dance material, or music and dance material, alongside each other.[9] We might be able to move from larger statements about the relations between artforms to relations between elements within a form.

We are much more able to accommodate juxtaposition in its linear manifestation, in the metonymic axis. It is when elements are piled up in the metaphoric or condensed axis that we begin to have cultural problems with their reception. This condensation relates to the next category of hybridity, synthesis, but could simply be dense juxtaposition in the moment. If we think of the example of cuisine, we are able to accommodate a great diversity of food during the linear course of a meal, but having all those things piled up in one course would outrage our sense of decorum, of good taste, and depending on our cultural conditioning, would provoke profound psycho-physical reactions: nausea perhaps. It is possible that the reactions to certain hybrid works is of a similarly psycho-physical nature. Certain possibilities of combination of art form products making us almost literally sick.[10]

In fashion, codes governing hybridity are less rigorous than the codes of cuisine. Cultural codes relating to art are closer to the freedoms of fashion, of course. The model of hybridity is paradigmatic, the codes of cuisine more syntagmatic. This is another of the taboos that hybridity and collaboration transgress. In hybridity, there is a move from the syntagmatic to the paradigmatic that unsettles the 'rightful' separation of elements (the enforcing of the status quo).

Whilst we can propose the clarity of the juxtaposition of image or object, there seems to be a strong force pulling towards two other conditions – synthesis and transformation. Juxtaposition creates shifting interpretative contexts – essentially transformative – for the various objects/images in the field under scrutiny.

## *Synthesis*

Synthesis operates when formal elements change in combination so that they cannot be separated. A concrete poem is meaning bearing text and visual construction. But both visual pattern and text are inextricable parts of the same thing. The work has no meaning other than when it is both visual pattern and text. Frank Stella's sculpture paintings are both flat surfaces with paint on – paintings – and they are also shaped, 3-dimensional and frequently situated in particular architectural situation – sculptures. But they are not separable. You can't take away the shaped surface to leave anything, the painting depends on that shaped surface and the effect of the shaped surface depends on the colour and texture. Both elements are there but they are inextricably connected. We cannot really talk of a sculpture and a painting, a poem and a pattern. In both those instances there is a new synthetic form made up of historically pre-existent forms (poem/picture, painting/sculpture) and a new hybrid form emerges, the concrete poem and the sculpture painting. Rauschenberg provides an example with his assemblages, and Jasper Johns with his collage paintings. It is clear that whilst considerations of music and dance push us towards considerations of synthetic works, there is a very real difference between a sculpture painting and a dance/music work of the Cunningham kind, for example.

Music and cinema may be synthesised where the film strip itself generates the sound by effecting the playback heads in the projector. A miked stage surface may not have any life apart from its activation by the steps of the dancer.[11] Sound and movement can only be fully realised in the moment of performance. Technology, then, can make art forms much more closely dependent on each other than previously, but close dependency is still not yet synthesis.

We can further imagine a synthetic form arising out of the work of more than one person in the manipulation of image and sound in the live event. A situation where sampling and mixing is reauthoring.

It is difficult to imagine a situation where music and dance are synthesised, unless we are looking at a dancer making his or her own music whilst dancing-the movement of the arms whilst playing a gong constituting the dance, say. Or we might be watching a musician integrated into the on-stage event where we are invited to read their functional movements as dance. We could think of Anne Teresa de Keersmaeker's *Stella* and Rinde Eckert with the Margaret Jenkins Company's *Shelf Life* where Eckert is mover, singer and action maker. But this situation impinges significantly on the following category, transformation.

## *Transformation*

Transformational hybrids operate when a work clearly within one historically defined arts form shows the profound influence of another, or where our perception of one art form is necessarily altered by its interpenetration, to some degree, by another art form. It is a case of one art form moving in the general direction of another. A kinetic sculpture is a sculpture moving in the direction of dance, but not actually dance. The primary art form remains clearly primary but with an inclination towards another, read in a significantly other way.

Anne Teresa de Keersmaeker's *Stella* is an interestingly complex case in point, to which I have already referred. It combines music, dance steps, action, text and stage architecture/design. At one point Ligeti's *Symphonic Poem for 100 Metronomes* is played, the metronomes being set off by the dancers who then leave the stage. One remains far upstage. Clad only in a towel she slowly moves between two stools, trying to keep the towel around her. A balancing act, equivocating between the two poles of the stools. During this the music plays itself out in real time. Here there is a complete piece of music. It is also the case that the sound and the action are contra standard for our expectations of music and dance. Metronomes are sound not music, holding a towel is movement or drama, not dance. There is no question of a face to face collaboration between de Keersmaeker and Ligeti, and yet here is the most uncompromising presentation of a musical work within the frame of the dance-work event which brings the relation of Kantor and Witkiewicz's works to mind. Here there is juxtaposition – two discreet art works existing side by side, dance upstage, music down stage, neither dependent for its integrity on the other. Here is transformation – the Ligeti, in the frame of the dance-work event, cannot but be re-read as movement itself, the arms of the metronomes articulating space and time. By extension, the movements of the dancer upstage become a kind of dividing, measuring activity, not a dance and not music. The forms seem to be reduced here to some basic principles, the division/articulation of space and time. They cannot but be seen through the frame of the other. What we take to be a synthetic event, the dance theatre event where music and movement are inextricable linked, we see a radical separation of the elements at the same moment when we can least tell them apart. And they might be considered transformational, the music, metaphorically, moving towards dance, the dance, metaphorically, moving towards music. Here is a kind of deconstruction of the dance event. Whereas we think of the dance event, usually as primarily dance with the support of music, de Keersmaeker shows us that both forms are linked by certain similarities of structure. That the two forms do not necessarily exist at two binary poles, but that there is the possibility that they could occupy the same conceptual and physical space. Her rhetorical strategy for effecting this deconstruction, however, this unsettling of established hierarchies, is to tip the focus in favour of the 'music' and away from the dance, but not simply reversing the hierarchy in favour of the music to the exclusion or detriment of the dance/movement.

What is interesting to me is that hybridity can be read as a metaphor for the creative and reading processes, where signifiers make sense by virtue of their places within systems, codes. Combining diverse systems (music, dance, painting, film) reframes signifiers allowing their meanings to mutate. The putting of things together to make new things, the transformations of things in relations to others. Hybrid art forms and the works they encompass tend to work as symbols of the act of creativity itself. Forcefully and purposively putting things together, of welding items previously disparate and unconnected into new and more complex unities. It also suggests that there can be no innocent juxtaposition, it always moves towards synthesis and transformation – reframing.

What is clear about the three kinds of hybridity discussed above is that not one of them suggests that collaboration – in the sense of a number of artists from different fields working together to produce the work – is a necessary condition of the hybrid work. We can also easily imagine a number of artists from the same discipline working together to produce a 'pure' artwork existing quite straightforwardly within one historically defined art form.

This brings to the fore the hidden assumption in collaborative work – however put together or initiated. Collaborations exist between artists from different artforms and the work they create will necessarily be hybrid works. However, what we have a focus on in the publicising and marketing of collaborations, is not the hybridity of the resulting artwork, but the glamour of the artists themselves. Collaboration can be a distinctly conservative gesture and the cultural disposition to autobiographise, to focus on personalities, to narrativise militates against the potential novelty (and difficulty) of the hybrid product. Personality neutralises difficulty of interpretation by effecting a reseparation of the elements back to the ownership of the individual. Hence a frequent mechanism is to read the product as being inscribed within a single signature.

The big issue we rarely deal with when talking about collaboration and the hybrid arts is that we make things with other people, other personalities. And we are not all balanced, rational, loving, generous people. We are driven, riven, emotional, twisted, neurotic, psychotic even. We are temperamental, we are moody, we are controlled by chemicals. The best intentions can be scuppered by one too many lagers the night before, bad potty training, or secretly held beliefs about life in general, other people, or political systems. Whatever the talk of equals, there's always some miserable fucker who wants it done another, usually their way. It reminds us that the other meaning of collaboration is colluding or co-operating with the enemy. How often the equal partner has become the enemy.

## *References*

1. In the current writing I look up the computer's thesaurus to generate the cognates of 'partnership': alliance, society, union, fellowship, association, brotherhood, coterie, fraternity.
2. In conversation during the launch of H.G. at Clink Street Vaults, 11 September 1995.
3. Related to me by David Vaughan, archivist with Cunningham Foundation, in conversation. It was apparently Cage who saw the costumes before Cunningham and suggested that they be changed before Cunningham saw them. Anecdotally, at least!
4. In Shyer, Laurence. *Robert Wilson and His Collaborators.* Theatre communications Group, New York, 1989. pp. 234–5.
5. Levindon in his extremely useful article (Levinson, J. 'Hybrid Art Forms', *Journal of Aesthetic Education.* Vol. 18:4, 1984. pp. 5–13), suggests these three kinds of hybrid art work.
6. *Robert Wilson: The Lecture,* Tuesday September 5th at London's Royal Geographical Society.
7. See my article, 'Screams in Hyperspace: Robert Wilson's Universe', *Dance Theatre Journal,* Vol. 10:2, pp. 41–43, where I suggest that the network of connections between images resembles hypertext, many reading and meaning, rich pathways passing through objects and images.
8. Sue MacLennan is an independent choreographer who lead a workshop and discussed her work during the '93 Exchange.
9. MacLennan notes that the dance frequently pre-exists the music and that they then develop in dialectical steps. But in many Classical and contemporary works the music pre-exists the dance and exists as a coherent piece in its own right, with the dance existing as a structuring of time in parallel. In this relationship the dance may bring out some of the emotional qualities of the music in another code, the expressive human code. But the dance could also exist separately, conceivably, even though based on the structure and emotional graph of the music. But it would not necessarily exist separately in the special sense of a Cunningham dance, which is conceived entirely on self-sufficient counts and spatial arrangement. One marker of its self-sufficiency is that it

could co-exist with any piece of music. The juxtaposition would change the resonances between all the elements, but the principle of reframing is one its premises. Of course, we could say that all dance could co-exist with any piece of music and that the reframing would change the meanings. Here we might have to make a statement about the historical or cultural placement of the dance. 'Swan Lake' could be danced to a rock score, and the meaning generated would have something to do with that breaking of a cultural unity. And there would be significant change in the technique necessary to sustain the dance. Rather than relying on the music to provide the emotional support, dynamics and counts, they might have to be internalised to sustain the parallel dance text. This might in turn affect the physical coding of inadvertent facial or muscular expression.

10. I might instance the mass and bad-humoured walkout of audience members at a performance of Anne Teresa de Keersmaeker's *Stella,* at Glasgow's Theatre Royal, during the section where Ligeti's *Symphonic Poem for 100 Metronomes* was 'played' alongside, but dominating, a minimal sequence of real-time movement activity. The metronomes which play out over some 15 minutes, alongside the minimal and non-technical dance was a juxtaposition which drove the audience to extremes of reaction. See Section III of this article for a more detailed account of Stella, and my article, 'Postmodernism Deconstructed', *Dance Theatre Journal,* Vol. 8:2, pp. 28–31.
11. I recall that the percussionist and composer Simon Limbtick (who was a Exchange participant in '93), made a piece, many years ago, where he sat at a desk which was miked up and played it percussively. His movements generated the sound of the piece and the combined effect was a synthesis of sound making and movement or action into a percussion dance. In terms of the signature, we seem here to be talking about an unqualified single signature on the work, whilst the work is synthesised from two or more pre-existing forms.

*Performing Arts International*
1996, Vol. 1, Part 1, pp. 21–30
Reprints available directly from the publisher
Photocopying permitted by license only



# Interdisciplinarity, Ideology and the 'I': Research Methodologies in the Performing Arts

Alexandra Carter

The identification of methodological approaches to an area of study, which serve to inform and guide analysis, is a key element in the formal research process. Examples of some of these approaches are extrapolated from research on images of women in the music hall ballets of late Victorian and Edwardian England. The potential of interdisciplinarity; the concept of ideology in relation to the production of knowledge by performance and by its attendant written and visual commentary, and the theories of feminism are offered as providing some pertinent methodological perspectives on gender and the performing arts.

KEY WORDS: Dance, gender, research, interdisciplinarity, ideology, feminist theories.

In scholarly research, the methodologies employed by the researcher are crucial, whether the outcome is of a written or practical nature. Harding (1987) offers a useful distinction between research 'methods', which she characterises as techniques of gathering material such as questionnaires, interviewing, statistical surveys, etc., and 'methodology' as the theories which inform procedure and analysis. That is, methodology is the conscious, articulated approach to research which helps form questions, guides the selection and interrogation of source material and has the potential to result in a new epistemological framework. In a formal thesis, this aspect of original scholarship is a key element.

The following exposition offers examples of some methodological approaches to research in the performing arts, with special reference to the study of how the gendered image is produced by the discourse of performance.[1] Although the focus is on the various perspectives employed rather than the content or outcome of the study itself, a brief outline is given of the parameters of the research topic. This is followed by a description of the three inter-related approaches or concepts which constituted the main methodological thrusts of the research. These were, firstly, *interdisciplinarity* which offered the necessary range of bodies of knowledge and concomitant theoretical concerns. Secondly, the notion of *ideology*, as embodied in existing knowledge and the 'new' knowledge of performance, was a central guiding concept for enquiry. Thirdly, the unifying theoretical framework for the research topic was adopted from feminist theories. These theories allowed the researcher to identify performance as a contributor to the social construction and regulation of gender and to claim the experiences of women in performance as constituting significant knowledge about the world. Furthermore, the intervention of the researcher, the 'I' of the production of that knowledge, is acknowledged as forming a necessary positional stance on the research.

It is important to stress, however, that this article exemplifies some possible perspectives but it does not preclude others. The researcher will identify the methodological framework most suited to her/his motivations, interests and subject matter. The subject matter of this case study was an investigation into how the image of the dancer was produced in the music hall ballets of Victorian and Edwardian England.[2]

Between 1884 and 1915, ballet was the main attraction on the programmes presented by London's Alhambra and Empire palaces of varieties in Leicester Square. At least one, usually two and sometimes three ballets were presented nightly at each venue and the productions normally ran for six months or more. Approximately one hundred and forty new works were created during this period.[3] The Alhambra and the Empire, which evolved from the traditional music hall, presented their ballets as part of a mixed or 'variety' programme. Although the choreographers, or arrangers and inventors as they were called, were both men and women, the works were produced to the dictates of an all-male management for a predominantly male audience. It is men's perceptions of the ballets and their executants which have been recorded in the written and iconographic discourses which constructed, and in secondary sources, continue to construct the image of the music hall dancer. This image was produced by the ballets themselves and by the power structures of their institutional context, and mediated in the 'factual' writings of journalism, biography, autobiography and history; the fiction of novels and poetry, and by the visual representation of the dancer in programmes, sketches and photographs. As none of the ballets are extant, it was these written and visual records which constituted the main source material for the study.

## *Interdisciplinarity, Ideology and the 'I'*

In order to analyse how the image of the dancer was produced, mediated and circulated, Wolff's contention that it is necessary to examine "both texts and artists within the complex system of aesthetic and ideological discourses, institutions and structures in which they reside" (1990, p. 112) was adopted as a guiding principle. Recent cultural theories claim that, in one sense, there is no distinction between text and context, between the artifact/performance and the circumstances of its production and reception (Appleby, Hunt & Jacob, 1994). However, to address both 'text' and 'context' is not to treat performances as discrete phenomena, for their meanings can only be experienced or read in relation to the 'ideological discourses' which produce them and are produced by them. To argue for an analytic framework which can be applied to the components and structures of the dance (or the music, or any theatre event) is not an attempt to fix or stabilise meanings; nor is it a claim for social, historical, artistic or any other kind of autonomy for performance.[4] On the contrary, it enables the researcher to identify the various constituent elements of a work(s) which exist for analysis from whatever perspective s/he chooses to employ. 'Text' and 'context' are contingent, and the following outline of methodological approaches demonstrates that contingency. These approaches endorse the argument that the performing arts are not just artistic products but are significant cultural events which contribute to the belief and value systems of a society.

In order to investigate how the image of the dancer was produced by "specific historical practices on the part of identifiable social groups in given conditions" and, furthermore, how this image bears "the imprint of the ideas, values and conditions of existence of those groups" (Wolff 1981, p. 49), it was necessary to draw upon a variety of disciplinary approaches. These approaches offered not only apparently discrete bodies of knowledge but also their distinct theoretical concerns. Analysis of dance performances of the past which were located in the realm of popular entertainment, and defined by the institutional structures and groupings which provided and received that entertainment, required perspectives drawn from history, cultural studies, sociology and, of-course, dance itself.

An examination of historical sources revealed the necessary information about the development of the music hall and its resultant characteristics. Set against the history of ballet and its concomitant image as an elite art form, the absence of the music hall ballet in historiography can be explained by a clash of two histories. The location of a 'high art' form in the context of popular entertainment resulted in the neglect, distortion or trivialisation of the era in both dance and music hall history.[5] As Chapman (1979/80) suggests in relation to dance, 'high' culture has been the focus for study but popular entertainment, seen as of no artistic or aesthetic value and even as morally corrupt, has rarely been documented. In the approaches of cultural studies, however, "leisure time and its activities are acknowledged as a significant element of social experience" (Bailey 1978, p. 1).[6] It is also the experience of the ordinary people of the past which has been neglected, for traditional history has been concerned with kings not commoners, with ballerinas not ballet girls,[7] yet as Hutcheon demands, we must now address "the lives of the losers as well as the winners ... of the unsung many as well as the much sung few" (1989, p. 66).

The approaches of cultural studies affirm the study of popular culture and place the phenomenological approach, that is, how people themselves think and feel about their experience and their place in the world, as central. As Barrett points out, these approaches deal not only with "those cultural products which have been rejected by the academic establishment" but also "the products can be studied alongside their consumers" (Barrett, *et al.* 1979, pp. 26/7). One of the concerns of the sociology of art is the relationship between the 'consumers' and the artistic processes and products of society, for "we cannot legislate about effectivity without reference to the specific circumstances of readers and viewers" (Wolff, 1990, p. 5). Neither entertainment nor the arts are isolated phenomena but they produce the values of society: "performances do not only reproduce or display ideologies but contribute to their construction and modification" (Bratton in Bratton, ed. 1986 p. xi). If it is recognised that all aspects of culture, including the arts, are ideologically situated, then the claim may be made for ballet as one of society's hegemonic practices. However, before any connection could be made between the ballets of the music hall period and the gender ideologies they (re)produced, it was necessary to examine their specificity as dance events. Although recent sociological debate has called for the integration of analyses of texts and institutions (see Wolff, 1990 as above), as a disciplinary stance it cannot provide the methodological framework for the analysis of the dances themselves.[8] For the researcher in the performing arts, particularly in the area of theatre, it is a challenge to devise systems whereby the components and structures of performance can be analysed as distinct from analysis of the score, script or written text.

Dance can be analysed from many perspectives. Most recently, psychoanalytic and linguistic theories have informed analysis, though there is the danger of the dance itself (or music, or theatre piece) disappearing in deconstructionist depths, constantly deferred meanings or fascination with the phallus.[9] Attempts have been made to formulate systems of dance analysis but these have tended to remain at the structural level of movement analysis. When expressed in notation, these systems have the advantage of relative exactitude but they are problematic in relation to the accommodation of the vital information which makes sense of movement. It is the model offered by Adshead (1988) which was used for analysis of the music hall ballets. This model embraces the relatively objective level of description which is directly related to the more discursive and complex notions of interpretation and evaluation. It is this comprehensive methodology which enables the components and structure of the dance to be identified and which makes sense of the meanings and significances of the works in their social, historical and artistic context.

To give a simple example from the music hall period, a descriptive analysis may reveal that the movement vocabulary of the *première danseuse* was based on the complex technical demands of the *danse d'école* and that of the *corps de ballet* on simple, spatial configurations and *tableaux*. How this difference is interpreted in its contemporary context clearly rests not only upon a knowledge of the history of dance traditions and dance hierarchies but also on an understanding of social stratification and career mobility in Victorian England. Another, more extended example demonstrates how sense can be made of the subject matter of the ballets, and roles of the dancers, in relation to the immediate location of their performance and to the wider, social context. An analysis of the narratives of the ballets reveals their relationship to the meta-narratives of society. The patriotic or military works reinforced concepts of nationalism and Empire; 'nature' ballets reflected neo-Romantic yearnings for the natural world; the inevitable happy endings endorsed the imperative of marriage and the danced roles of the ballerina and the ballet girl produced binary constructions of femininities which accorded with the sexual psyche of the era. Furthermore, an examination of the palaces of varieties, their audiences and the commercial imperative which guided their programming and their productions reveals the role of these venues in mediating the ideological imperatives of the period. On first consideration, the role of dancer-as-flower may appear to have little to do with anything except ballet tradition but by locating this image at the centre of the matrix of circumstances which produced the image, it can be seen as a direct manifestation of many of the sexual and psychological concerns of the age.

Overall, by drawing upon the various disciplinary perspectives outlined above, the study attempted to create a convergence of disciplines and, as Butler (1990, p. xiii) expressed in relation to her own work, sought "to affirm those possibilities on the critical boundaries of disciplinary life." The aim was to consider not only how various disciplines reveal different perspectives on the social construction of gender but to inter-relate those perspectives. The purpose of such an inter-relationship was not the achievement of a forced synthesis or a neat and tidy package of knowledge, but for the purpose of exploring both mutuality and difference. Barrett suggests that interdisciplinarity is not a search for a 'general theory' or an 'integrated perspective' but

> "on the contrary, to address the specific 'truths' of the different disciplines is to discover not the controlling modernism of a fully integrated theory of knowledge, but precisely the reverse – an incommensurability of knowledge that provokes interesting reflection."
>
> Barrett in Barrett & Philips, 1992 pp. 214–5

By examining the representation of dancers within the symbolic constructs of the ballets, exposing the relationship between their identity as performers in these ballets and their own, lived experience of the world, and reading both in relation to the historical, social and cultural context, the critical boundaries of dance analysis, history, cultural studies and sociology were merged, but their incommensurability retained.

Research, of whatever nature, relies on probing many different kinds of source material. The post-modern imperative to expose the ideologies embodied in the production and transmission of knowledge provides the opportunity (or necessity) for a critical analysis of how source material contributes to the construction of 'facts', opinions, beliefs and values. Such an activity does not dismiss or devalue the documentation but demands attention to the perspectives from which it was created. Any notion of historiographic neutrality is called into question, for example, by the previously mentioned omission of events or people from published records. However, the challenge made by contemporary cultural theorists that there can be no possible authorial 'objectivity' at all is counterbalanced by Appleby *et al.* (1994, p. 254). They offer a reworked notion of the term 'objectivity' which "conceedes the impossibility of any research being neutral" but which affirms "the viability of stable bodies of knowledge that can be communicated, built upon and subject to testing." Therefore, exposure of the subjectivity of authors, or other makers of knowledge, does not preclude 'objective' testable hypothesis. It does, however, reveal how knowledge is constructed and invites the researcher to speculate on the usually unrevealed perspective and purposes embodied in sources.

Ideology is produced not only by documentation but also by performance itself. As discussed above, it was an analysis of the ballets which exposed their hegemonic function in society. Analysis of how the image of the performer is constructed also has the potential to reveal how the identity of the performer can become conflated with the identity of the person as private individual. For example, the received image of the music hall dancer, particularly the ballet girl, was one of a woman who was at best, a dilettante who colluded in the flirtation across the footlights demanded by the context of her work, or at worst, an immoral figure who engaged in more than flirtation with the male patrons. Like the actress, by positioning herself as public property on the stage, her private morality was called into question.[10] In order to challenge this image, or test its tenability, it was necessary to expose the ideologies of sexuality and morality which were (re)produced in the written, iconographic and choreographic source material. Whilst exposing how knowledge about performance and performers is constructed by ideology, it was apt to acknowledge the stance which motivated my research; that is, my own authorial bias. The approach which allowed for such a declaration of bias as a strength, which owes its allegiance to no single disciplinary stance and, therefore, is able to accommodate the complexity of such a concept as the gendered image, is that of feminism.

The fundamental concern of the research on the image of the dancer in the late nineteenth century was an examination by a woman, of an activity undertaken by women, in order to understand how perception of that activity has been influenced by patriarchal ideologies both then and now. Traditional disciplinary theories provided the necessary but not the sufficient conditions for such a study, for history has been notoriously biased towards the history of men and cultural studies, until recently, tended towards the male 'actors' experience. Similarly, sociological approaches, whether structuralist, phenomenologist or Marxist also tended to ignore the women's role and their perceptions of that role. Harding (1987) suggests that knowledge is only legitimated if it is recorded by men or is about men's experience and that

> "traditional theories have been applied in ways that make it difficult to understand women's participation in social life or to understand men's activities as gendered."
>
> Harding, 1987 p. 3

There is a distinction to be made between the study of women in history and a feminist approach to that study. A distinctive feature of a feminist approach is that it embraces methodology as well as subject matter. It not only acknowledges women as equally valid subjects for study but raises fundamental questions as to *how* such study can be approached. However, no one single theoretical paradigm for the study of women in history is offered, for "no one method, form of writing, speaking position, mode of argument can act as representative model or ideal for feminist theory." (Gross in Pateman and Gross, 1986, p. 204). As there is no one 'feminism', there is no single emanating theory. However, theory can be characterised, if not defined. Gross (Pateman and Gross, 1986) offers an outline of significant characteristics as (a) a recognition of the patriarchal values which dominate both traditional discourse and its application; (b) a recognition of the gaps and silences in discourse; (c) an examination of the reasons for those gaps and silences and; (d) the development of other modes of research even if it means using patriarchal texts as a starting point.

The importance of these ideas for the study of how images of dancers were created is evident when the ballets, and the written and visual sources on the ballet, are examined. Produced primarily by the men, they reflect the values by which women and women's bodies were defined. The gaps in both dance and music hall historiography, and the deep silence which surrounds the life and experience of the thousands of women who contributed to this significant form of cultural activity, are worthy of examination. Instead of accepting the gaps and silences as an indication that the period is not worthy of attention and the contribution of the people as not worthy of record, the researcher needs to be alert to what is missing from history, for "historical materials are preserved when it is in someone's interest to preserve them" (Smith in Carroll, 1976, p. 372). If there are no records, other texts have to be probed for clues to unrecorded experience. Like so many performing artists who never achieved stardom, notoriety or public attention, the lived experiences of the *corps de ballet* can be explored through "deduction, inference ... and ... symptomatic reading" (Allen in Pateman and Gross, 1986, p. 176).

The research discussed in this exposition was informed by the general characteristics of feminist theories as outlined above. However, the particular concern was with the relationship between the performing arts and gender, most specifically how body image was gendered in the (re)presentation of the dancer on stage. Areas of debate therefore cohered around general notions of how gender is produced, and the specific notion of how gender is produced by the arts.

Early feminist debate tended to treate the concepts of sex as biologically determined and gender as socially constructed as unproblematic but, more recent writers such as Butler (1990) raise questions in relation to the fixed or free will nature of these concepts and their complex inter-relationship. The difficulties of examining the relationship between sex and gender are compounded in dance, where the gendered activity is located in the body itself. In order to examine how dance constructs gender, Butler's characterisation of gender not as

> "construed as a stable identity or locus of agency from which various acts follow – rather, gender is an identity tenuously constituted, instituted in an exterior space, through a stylised repetition of acts."
>
> Butler, 1990, pp. 40–41

Dance is clearly a 'stylised repetition of acts', albeit far more formal and self-conscious than the everyday activities of life which, perceived as natural, blur the social construction and regulation of gender.

## *Conclusion*

The relationship between gender as performative act and dance as performing art is more than one of alliteration. As de Lauretis (1984, p. 38) claims, an examination of the "representation of the female body ... necessarily constitutes a starting point for any understanding of sexual difference." Although *what* the body signifies to a spectator will vary according to factors such as the social and historical context of its presentation, and the identity and personal history of the spectator, the notion that "the cultural significance of the female body is not only ...

of a flesh-and-blood entity, but that of a symbolic construct'' (Suleiman, 1986, p. 2) is central to a feminist approach to the arts. In its attention to what the female body symbolises and how it comes to do so, feminism has the potential to offer a radical subversion of traditional readings of the arts.

Furthermore, it challenges the conventional divisions between art and popular culture, claiming both as practices which produce dominant concepts of femininity. It also challenges the compartmentalisation of knowledge, which prohibits the kinds of comparative study which can expose the pervasiveness and mutually reinforcing nature of these concepts. In challenging the supposed neutralities of knowledge and the producers and mediators of knowledge, of the languages of discourse and particularly the symbolic languages of performance, it exposes its own bias. A feminist researcher is one who seeks to validate the experiences of women from their own perspective as constituting significant knowledge about the world, and exposes her own motivation for doing so. The intervention of the 'I', in an interdisciplinary approach to research, reveals the ideological constructs of femininity which are embodied in the performing arts not as predestined, inevitable or immutable, but as hegemonic systems of power and control.

## Notes

1. The term 'discourse' as used here is characterised by Scott (1988 p. 35) as 'not a language or a text but a historically, socially and institutionally specific structure of statements, terms, categories and beliefs.' Her contention that 'discursive fields overlap, influence and compete with one another' is central to any analysis of performance as artistic, social, historical and, as claimed in this article, ideological discourse.
2. The unpublished thesis, entitled 'Winged and shivering: images of dancers in the Alhambra and Empire ballets 1884–1915' was awarded a Ph.D. by the University of Surrey in 1993.
3. See Guest (1992) for a detailed account and choreochronicle of the works presented at the Alhambra and Empire palaces of varieties.
4. See Shepherd (1991) for a detailed argument claiming music as a social, rather than just an autonomously 'musical' text.
5. Haskell (1938), Lynham (1947) and Clarke and Crisp (1981) exemplify how the period has been overlooked for serious attention in secondary sources on dance history. A typical example of neglect in music hall history is Green's anthology (ed. 1986) which has a colourful front cover depicting an Empire programme featuring its *maitresse de ballet,* and contains many valuable references on their experiences of the ballet from writers such as George Bernard Shaw, Arnold Benner and A.J. Symons. Curiously, neither 'dance' nor 'ballet' are menitoned in the lengthy index which is often the first consultative point for the reader/researcher.
6. At a recent study of Dance conference, it was evident that dance scholars have expanded their parameters not only to embrace difference methodological frameworks but are also now attending to a wide range of functions and forms of dance, including those which

constitute leisure activities. (Proceedings, *Border Tensions: dance and discourse* National Resource Centre for Dance at University of Surrey, 1995)

7. Whatever their age or marital status, the dancers of the *corps de ballet* were described as 'ballet girls.'

8. A systematic approach to the music hall ballets was particularly pertinent as none are extant and little is known about their characteristics.

9. See Adshead-Lansdale (1993/4) for her concern about the wholesale adoption of theories from other 'disciplinary' perspectives, particularly literature and cultural studies, which undermine the status and richness of dance as a field of study in its own right.

10. See Davis (1991) for an analysis of the sexual identity of the actress in Victorian England.

## References

Adshead, J. (1988) Editor. *Dance analysis: theory and practice* London: Dance Books.

Adshead-Lansdale, J. (1993/4) Dance and critical debate: towards a community of dance intellectuals. *Dance Theatre Journal* (11/2) pp. 22–24; 33.

Appleby, J., Hunt, L. & Jacob, M (1994) *Telling the truth about history* London: W.W. Norton.

Bailey, P. (1978) *Leisure and class in Victorian England* London: Routledge Kegan Paul.

Barrett, M., Corrigan, P., Kuhn, A., Wolff, J. (1979) *Ideology and cultural production* London: Croom Helm.

Barrett, M. & Phillips, A. (1992) *Destabilising theory: contemporary feminist debates* Cambridge: Polity Press.

Bratton, J.S.. (1986) Editor. *Music hall: performance and style* Milton Keynes: Open University Press.

Butler, J. (1990) *Gender trouble: feminism and the subversion of identity* London: Routledge.

Carroll, B. (1976) Editor. *Liberating women's history: theoretical and critical essays* Urbana: Univ. of Illinois Press.

Chapman, J. (1979–80) The aesthetic interpretation of dance history. *Dance Chronicle* (3:3) pp. 254–274.

Clarke, M & Crisp, C. (1981) *The history of Dance* London: Orbis Publishing.

Davis, T.C. (1991) *Actresses as working women: their social identity in Victorian culture* London: Routledge.

de Lauretis, T. (1984) *Alice doesn't: feminism, semiotics, cinema* London: Macmillan.

Green, B. (1986) Editor. *The last Empires: a music hall companion* London: Pavilion Books.

Guest, I. (1992) *Ballet in Leicester Square* London: Dance Books.

Harding, S. (1987) Editor. *Feminism and methodology* Indiana: Indiana University Press.

Haskell, A.(1938) *Ballet* Harmondsworth: Pelican.

Hutcheon, L. (1989) *The politics of postmodernism* London: Routledge.

Lynham, D. (1947) *Ballet then and now* London: Sylvan Press.

Pateman, C. & Gross, E. (1986) Editors *Feminist challenges: social and political theory* London: Allen & Unwin.

Scott, J. (1988) Deconstructing equality – versus – difference: or, the uses of poststructuralist theory for feminism. *Feminist Studies* (14/1) p. 35.

Shepherd, J. (1991) *Music as social text* Cambridge: Polity Press.

Suleiman, S.R. (1986) *The female body in Western culture: contemporary perspectives* London: Harvard University Press.

Wolff, J. (1981) *The social production of art* London: Macmillan.

Wolff, J. (1990) *Feminine sentences: essays on women and culture* Cambridge: Polity Press.

*Performing Arts International*
1996, Vol. 1, Part 1, pp. 31–51
Reprints available directly from the publisher
Photocopying permitted by license only



# Architecture and Dance

Emmerich Simoncsics, Evelin Gruber, Andrea Kocevar, Michael Fiala, Renate Hammer, and Peter Parolek

*Applied Aesthetics Laboratory, Tokyo University.*

An account of theoretical and practical exploration running across disciplinary boundaries undertaken by the Applied Aesthetics Project, centred at the University of Tokyo and involving also the Technical University of Vienna. Architecture students were asked to produce schematic designs in response to their observations of dance; the outcomes suggested some surprising and liberating new design concepts, bringing together two areas of spatial, dynamic art and revealing unexpected aspects of commonality.

KEY WORDS: Kinaesthetic sense, dance positions, architectonics, rhythmic gymnastics, Bauhaus.

For the dancer an economical interplay of forces in the human body is essential. Likewise for the architect it is important to find the statically correct structural form of a building. It is the kinaesthetic sense that enables the designer to do this; without it the construction is not only uneconomical, but also illogical. Dance is movement, dance positions are an integral part of movement – they are a momentary cut in the course of dance. Dance positions have a striking visual impact and can inform analysis of spatial features and proportions, as well as suggesting transformation into another medium, such as architecture.

In experiments described in this article, identification of relevant dance positions for architects was carried out in cooperation with dancers. Typical structural forms were compared with dance positions; similarities and differences were noted. Test students of architecture analysed specific positions and drew the interplay of forces in the human body. It turned out that some positions had interesting force transmissions, which had not yet been fully realized as a possible construction system for architecture. Having had this experience, dancers also began to develop new positions suitable for transformation into architecture.

*Performing Arts International*
1996, Vol. 1, Part 1, pp. 81–86
Reprints available directly from the publisher
Photocopying permitted by license only



# Depression in the Land of Dance Administrators

Julia Carruthers

What do dance administrators moan about when they get together? Why is the mood generally so despondent and weary? A brief look at self-esteem, the response from promoters, the use of meagre resources, and the lack of recognition by one who has been crazy enough to rejoin the club.

KEYWORDS: Dance administration, administrators morale, Arts Council funding, English and American dance administration.

Suzanne Walker, the administrator of V-TOL Dance Company and I are sitting in a crowded pub garden in Clapham. It is a beautiful, hot evening and there is a magnificent sunset starting. We are exchanging heart-warmingly disloyal and bitchy stories about the land of dance and dancers and we are laughing a lot.

"Well, you seem all right," I say, "Here I am trying to collect evidence to support the much repeated maxim that all dance administrators are depressed and you're sitting there looking tanned and healthy with a big grin on your face... I thought dance administrators were all supposed to have anxiously hunched shoulders, mumble wearily into their complimentary bottles of Beck's beer and constantly shake their heads."

At a recent meeting of the Management Liaison Group, a regular talkshop for administrators of small and middle-scale dance companies, Jane Attenborough (Director of Dance UK, the UK's umbrella organisation for Dance) offered to invite a psychologist to the next session, as she was worried about the dreadfully despondent mood amongst the group. She was not joking.

I ask Suzanne what she thinks some of the problems are. A deep sigh and slumping of the torso follows (you understand that dance people follow body language closely). She pulls a rumpled notebook out of her bag and flicks efficiently to where she has scribbled her findings. She has consulted with other dance administrator colleagues, having been

Hauptmann and many others. They admired the unity of music, movement and light, which was seen there for the first time. Several students of Dalcroze and the subsequent school in Laxenburg, Austria found their own way and language for dance, gymnastics and music. Some of them (for example Rosalia Chladek, Mary Wigman, Christine Bair-Frisell, Valeria Kratina, Suzanne Perottet) influenced the course of dance history and taught at the school after their studies.

In 1919, the architect Walter Gropius assembled a group of avant-garde artists in Weimar, Germany (including Johannes Itten, Lyonel Feininger, Paul Klee, Oskar Schlemmer, Wassily Kandinsky, Laszlo Moholy-Nagy, Gerhard Marcks, Georg Muche, and Lothar Schreyer) in order to set up the Bauhaus. Their goal was to depict man in his entirety in the broadest sense. In 1929, as Schlemmer left the Bauhaus, he spoke about stage events as a forum for the metaphysical, as opposed to purely representational aspects. As early as 1912 he started to draft a plan for dance with the dancers Albert Burger and Elsa Holzel. At this time Schlemmer studied in the masterclass of Adolf Holzel. Through their studies the two dancers were influenced by Dalcroze. The result of Schlemmer's plan was the *Triadische Ballet*, which had its premiere in 1922. The name of this ballet without plot was derived from the Greek term triad, the organizing principle, from which the dance was created. Schlemmer called it an "apotheosis of the three", the change of one, two and three in shape, colour and movement. These factors were also decisive concepts underlying the dancers' transformation in costumes (see Oberzaucher-Schuler, 1992). Later on he created some dance/mime études, which were also performed on the Bauhaus stage. Human presence was not a portrayal of individual expression in these dances, but – standardized by costumes – was intended to be a manifestation of the formal stage elements: space, form, colour, light, and materials. The dancer reacted as a body mechanism or reacted to equipment and requisites (ibid. p. 307).

Kandinsky fulfilled his synthesis of dance and painting for the first time in *The Yellow Sound*, and later in Mussorgsky's *Pictures at an Exhibition*. The premiere of *The Yellow Sound* took place on the Bauhaus stage in 1928. Kandinsky cooperated often with the dancer Gret Palucca, who also repeatedly performed on the Bauhaus stage. Kandinsky was fascinated by the acrobatic mastery of her physical expression and positions which strengthened the spatial concept he sought. He captured several moments of her with graphical diagrams.

During the 1920s in Paris, dancers and choreographers collaborated with painters and composers. There the Russian Sergei Diaghilev led the *Ballets Russes* and Igor Stravinsky composed pieces for the company, such as *Petruschka, L' Oiseau de Feu, L'Après-midi d'un Faune* and *Le Sacre du Printemps. L'Oiseau de Feu* helped Stravinsky make his breakthrough as an international success, while *Le Sacre du Printemps* encouraged a

Three dimensions: drawing by Andrea Kocevar.

tumult after a few bars and caused one of the best known scandals in musical history. Vaslav Nijinsky also found his artistic acceptance as a choreographer in the company. In *L' Après-midi d'un Faune,* his first choreography, the dancers moved along a line parallel to the audience – the dimensional possibilities of the dancers were comparable with a Cubist painting, projected onto a plane. In 1926 Joan Miro designed the costumes for the ballet *Romeo and Juliet.* Other painters such as Pablo Picasso, Henri Matisse, George Braque, Giorgio De Chirico, Max Ernst and the composers Maurice Ravel, Claude Debussy, Manuel de Falla, Eric Satie, Sergei Prokofiev also worked for the *Ballets Russes.*

In Italy in 1911 the dancer Rudolf von Laban made the first attempt to create an artists' association and to set up open air festivals. During the winter he moved back to Munich, Germany to continue his work. In Munich he also opened his first school. He assimilated the revolution in movement initiated by Isadora Duncan and developed it further in his book *Choreutik* (first published in 1966 although it was written at the end of the 1930's). This theory of movement is still one of the most important theoretical approaches to Modern Dance. For Laban, space was basically in motion, and each body forms a continuum with it.

"Yet we may not simply regard the place as an empty space separated from movement, nor may we see movement as no more than occasional occurence, for movement is a continuous flow within the space itself, and this is the fundamental aspect of space. Space is the concealed basic feature of movement, and movement is a visible aspect of space"

(Laban, 1966 p. 19).

the supposedly quaint idea that women are not used to putting themselves first. The irony is that the dance administrators may have to give up on the job once they have children, as the cost of childcare bears no rational relationship to the wages they are earning. This is just one of the worries about career development.

In the land that really gave birth to contemporary dance, America, administrators get called 'Executive Director'. They 'talk up' the 'product' they represent and its importance. In England the word 'administrator' implies that the people running the companies are secretaries. We English working in dance are modest and, comparatively, inarticulate. We're too shy and polite and we certainly haven't got all that American energy. A delegation of American dance administrators from companies such as Urban Bush Women, Bebe Miller and Doug Elkins visited the UK recently and met with a peer group of English dance administrators. The English crowd felt depressed and cowed by the dash and animation of the Americans. It wasn't actually as if the quality of the work was objectively superior or that the companies were doing better financially, but the Americans seemed to feel so much more healthy and optimistic about what they were doing (or they sure put on a damn good show and should be giving us acting lessons!). Of course, we must remember that in the United States someone working behind the counter at a car rental company will make the task of hiring you a car seem vital to the future of the human race and its sanity. This is because the whole psychology of America is that there is a pot of gold at the end of the rainbow. It appears to be too hard for the UK dance administrators to believe in that rainbow very often.

Those working elsewhere in dance or maybe on the Boards of dance companies and observing the administrators in action are liable make more general comments and possibly have an cooler-headed view:

> "The trouble is that the administrator is left to cope with everything. They are always where the buck stops. Some of these dancers and choreographers are so helpless. The administrator is left keeping the Board, the dancers, the technical team, the composer, the sacked designer, the journalists, the photographer, the minibus driver and goodness knows who else happy. I've seen situations where the administrator has to cope with a smoke machine not working in a theatre 150 miles away from the office, because the artistic director and technician are incapable of taking constructive, speedy action. Or suddenly the lonely administrator is telephoned from Tokyo because the artistic director has overslept, missed his flight and wants his office to sort it out. No wonder these people get fed up and burnt out. Don't forget as well that the artists can enjoy the refreshment of funded "Research and Development" time, which the administrators don't get!"

I interviewed another administrator who I caught red-handed scanning the pages of The Guardian on a Monday (for the Creative and Media section, in other words the arts administration jobs). She was anxious to remain anonymous. Her summary of the issues was becoming familiar:

"The problem is that the same old problems keep coming up and it makes you weary. However advanced you get, you never stop having to do the mundane tasks. I consider myself pretty exprienced and quite a sophisticated operator, yet I still feel abominably patronised by many of the people I deal with. Maybe it's because I earn much less than they do and because I work in the artform that hasn't been allowed to grow up yet, called Dance. You are never able to keep up with the workload. It's like one of those dreams in which you are running and running to keep up and it's all hopeless. I often have fantasies about working for a Festival, because I don't mind working my guts out, but it would be nice to think it would all be over in three weeks! I used to want to work for a funding body to have a rest from it all and the pension, the expenses, the maternity leave were appealing – but it now seems far less attractive – all the funding officers I know keep complaining about filling in charts with targets and ticking boxes."

A comment from one of the funding officers was sympathetic:

"Although the opportunities for networking and meeting have improved tremendously, the smaller-scale dance administrators do all lead such isolated lives. They are rarely able to learn from anyone in a better position. Morale is buoyant enough on the large-scale and people still seem to have a sense of humour there, but the small and middle-scale workers have very little contact with the General Managers of the biggest companies. I think that the large-scale administrators are paid a lot more, but actually don't have that much more responsibility, they aren't carrying that much more weight. The worries are essentially the same."

Many of the dance administrators arrive in the jobs with little previous experience. They are young and hungry and are perhaps willing to earn a lower salary. There are few assistant administrator jobs in the dance world in which people can learn the ropes safely. There are many baptisms of fire. The fast track can be exhilarating, but suddenly the tarmac can run out and the road becomes awfully bumpy. People tend to stay in the jobs too long, as there are no obvious next steps. They are more likely to feel stuck nowadays, as companies rarely develop resources quickly and quantum leaps in public subsidy are virtually impossible.

Back to Suzanne and I in that bar in Clapham. The sun has now set. It is getting cold. We decide that we are working in dance because choreographers and dancers are some of the nicest, most wonderful people we have ever met (and a darn sight more congenial to work with than theatre directors, actors and musicians!). We wonder what some solutions might be. A summarised list of our curtailed discussion (we have no sweaters to put on, we have not been out with a wollen article of clothing for weeks) might run as follows. More dance links with the Independent Theatre Council (which provides excellent support to small-scale touring companies), better pay (somehow!), sabbaticals, better links with the large scale companies, offices in the same buildings as rehearsal studios, more staff in the office, a huge generic marketing interpret the various heights in compositional opposites, for instance the heaviness of the floor level compared to the lightness of the tip-toe. The vertical development can also be taken up in the echelon formation of the floors. The still position in dance, similar to the pause in music, cannot be equated with stagnation. The still position is part of the movement, just as the pause as a non-sound is a proportion to the sound as an event, and thus part of the music. The style of the choreography as a whole and the dynamics of a frozen movement can be perceived from the dance position as an instantaneous photograph.

These components are essential in giving form and content to the architectonic design. For the required analysis of the inspirational model the designer must consider various architectonic elements simultaneously:

- the composition of the elements;
- the function of the building and the rooms;
- the selection of the materials;
- massive or skeletal construction;
- transparent or closed space.

The space in between and the three-dimensional quality of the dancers can create sequences of different rooms, which were arranged in one composition. Decisive here is the individual fantasy and the degree of individual abstraction. These points determine the quality of each project, so that with the same dance position, new ideas will always evolve. The architectonic result displays the degree of analysis and the ability of a free and playful interpretation as well.

Diverse aspects of the model used as inspiration can be formally transformed into architecture, for instance:

- the proportions of the dance position;
- the spatial relation of the dancers in relation to one another;
- the transmission of force;
- the expression of the dance position.

In the creative process it is possible to vary aspects of the dance position, for instance by:

- repeating the dance positions;
- changing the spatial order of the dancers;
- combining different scales;
- reduction of the linear system.

The human body, like architecture, reacts to gravity. The dancer is able to counteract this with his/her varying energy expenditure. In order to stay in a balanced position, it is necessary for the dancer to equalize forces through minimal movements. From the point of view of a constructor the dancer operates mainly with labile states of equilibrium.

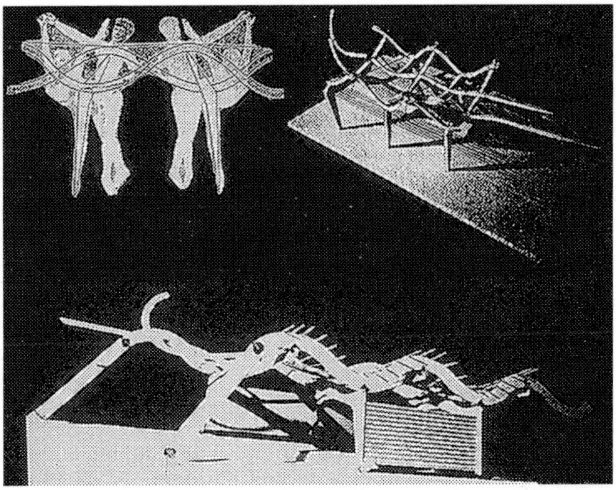

Railway Station/Thomas Mayer. The idea to shift the positions leads to a clear constructive solution: each one is carrying the load of the other.

In architecture, the construction and the resistance of material build a system in a stable state of equilibrium to direct the quantifiable load onto the ground. The depiction of transmission of force in the model requires a creative process from the designer. To reduce a dance position to a stable system of architecture, the essential expression of the forces of the dancers must be realized.

The range of transmission of force between two or more dancers ranges from slight touch to an embrace. These contacts are as important for the static connections in architecture as for the expressiveness of dance. The context of a choreography creates connections between the dancers, which are comparable with ideal force transmissions in architecture. The tension thereby created demands no body contact. Space is nevertheless created this way. The effect of these spaces of tension allows for many interpretations. All movements are a never ending change of connecting and loosening. Stability in dance means neither a complete stillness, nor an absolute stagnation – it tends to depict temporary, relative peace to achieve balance. Opposing this means mobility, a tendency to depict lively, fluent movement. This leads to a temporary loss of balance (Laban, 1991, p. 98).

The aesthetics of a movement and its expression can be transformed into architecture by a formal approach. Lightness, as a quality of movement, can be visualized in the dynamics of architectonic form. It is possible to transform it through:

- an extended projectile;
- a structure on a small supporting surface with minimal supporting points;
- loads, which are carried by light construction elements to a great height.

For this discussion it is assumed that the club goers are heterosexual and I do not differentiate between the behaviours of different ethnic groups, although there is an implicit examination of a certain class of youth by the nature of the venues that were studied. Youth, in this study, means men and women between 18 and 25, officially, and unofficially as young as 15 or 16 years old.

## *The social dance event as a mating ground*

Social dance spaces from dancehalls to clubs have been considered to be places to meet new people or, more specifically, new partners. Nightclubs are described by the users as 'cattle markets', 'pick-up joints', 'pulling places' and so on and Mungham (1976) aptly describes them as places where 'youth can pursue itself'.

Tracing the historical development of social dance since 1945, it is clear to see the move away from partnered and set dances to individual and improvised dances – a trend which began in the 1960s as part of the ethos of the time. Although there were some partnered dances as part of the 'disco' era in the 1970s (for instance the Spanish Hussle), since the 1960s, few set/partner dances have been prevalent and none appear on the dance-floors in the early 1990s.

Since my original research, several Latin-American dance crazes have had an influence on urban dance-floors. Certain venues have a regular Salsa night where a two hour class is followed by an evening of Salsa dancing. I maintain, however, that people dancing in the mainstream suburban nightclubs do not dance with a partner.

## *How does dance work as a courting mechanism in suburban nightclubs and speciality partner dance venues?*

My findings showed very clearly that women danced a great deal more than men in suburban nightclubs and that women were considered better dancers than men. The statistics I collected show that two-thirds of nightclub attendees are men and that two-thirds of the people on the dance-floor are women. This immediately sets up the familiar dynamic of the nightclub where the majority of men watch the women dance.

There are many well rehearsed arguments about why men don't dance (dance as a 'feminine art form/activity because it is both expressive and body centred) which I will not examine here. I see indications of this image of dance changing. There is now a much greater interest in the male physique shown in advertising and in men's magazines, the partner dance crazes, male roles models dancing (in the Guinness advertisement and the credits to the television series *Men Behaving Badly*) and the explosion of all male bands who dance as an intrinsic part of their performance (Take That and so on).

Whilst questionnaire data refutes that the women dance for the men's approval, interview data and observation suggest otherwise. The interviewees stated that they made more effort to dance well if they knew someone, especially if it was a man, was watching them. There is a limit, however, to what you can do to attract a non-dancing potential partner.

All the available data suggests that women are much less likely to initiate contact with a non-dancing, potential partner than men. Once again there are well rehearsed reasons for this, to do with the continuing stigmas against women who show 'too much' interest in men – 'slags'. The women interviewed stated the various means by which they would attempt to attract attention from potential partners who were dancing, but the majority agreed that they would not use a direct, verbal approach to either dancing or non-dancing potential partners.

The two main strategies used to attract a man's attention whilst dancing are as follows:

1 Copying or responding to the other person's movements. This is used as part of a general indication of friendship and can be clearly seen among groups of friends dancing. It is good as an initial link. It shows observation, sensitivity and interest in the other person. What it will not indicate is any sexual interest as it is recognised as being a friendly exchange.

2 Dancing close to someone, accidentally touching someone or bumping into them.

This is clearly a more flirtatious strategy involving actual (or near) body contact. Interviewees stated that this would only occur after a period of copying/responding or after other indications that the male counterpart was interested. Observation shows that a crowded dance-floor gives quite a bit of opportunity for this courting strategy to be explored.

Although these two dancing strategies have been expressed by a number of interviewees, the majority of experiences indicate that it is not until eye contact is established that any indication of interest can be registered. In this way, the non-dancing potential partners are still able to partake in the courtship activity by watching a dancer that they like and waiting for her to 'catch their eye'. Repeated glances are an indication of interest and are used as a prelude to verbal contact which is usually in the form of asking the woman 'to dance' or to have a drink.

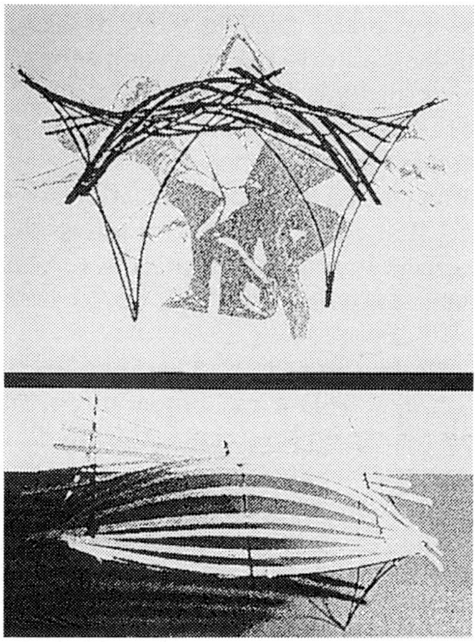

Ravilion/Eckhard Meyer. All analytic drawings of dance are by Andrea Kocevar.

## *Some specific experiments:*

**(1)** ***Transformation of movements describing space***

A different approach determined one experiment, which sprang from an exploration of space performed without music for architecture students in order to transform it into a composition of architectural space. Because of the ephemeral character of dance, the designer must keep the whole dance in his mind during the experiment, so in order to be able to transform the dance into architecture it was shown several times. Under these circumstances the designer was able to observe the dancers and analyse their proportions. Students perceived each position with their kinaesthetic sense and translated the force transmissions and the shifts of weight into a structural system. Following the space-describing transitions from position to position, and turning them into architectural space, the dance as an entire composition had finally to be transformed into an architectural composition, a process that demanded a high degree of abstraction and creativity (*Renate Hammer*).

**(2)** ***Transformation of movements describing solids***

In another project, the dimensions of dance were presented as additional information to a picture or a painting. A picture is two dimensional information – even if it shows a perspective (or an axonometric) view of an

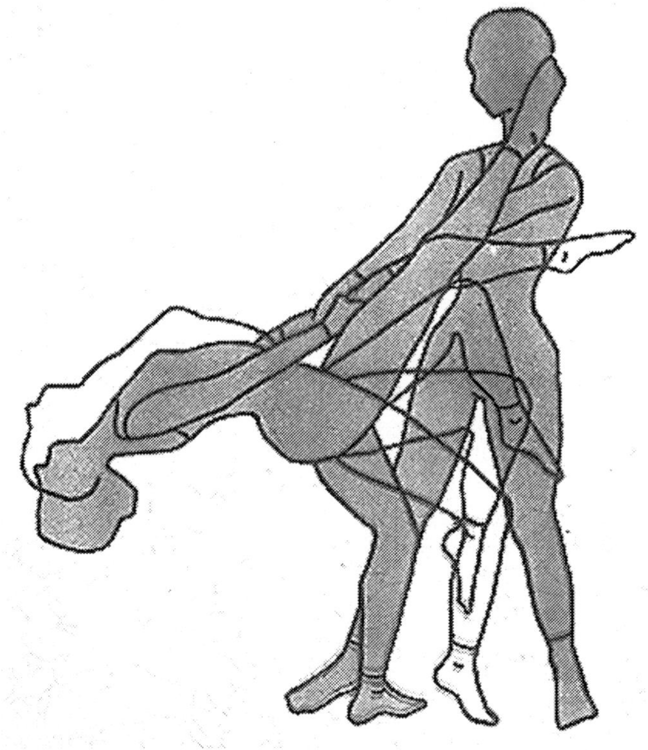

Drawing by Renate Hammer.

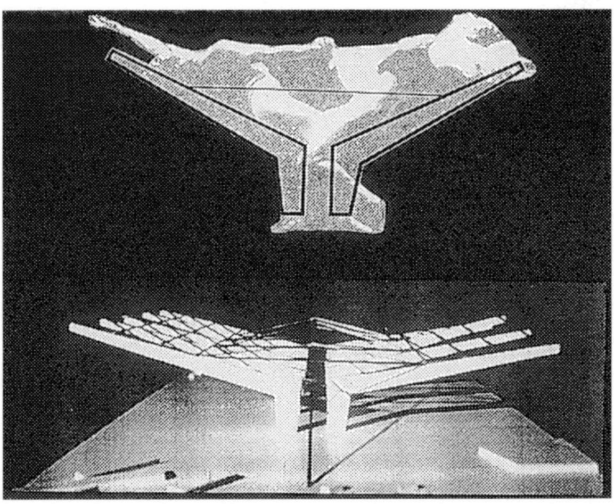

Landing Moorings/Seamus O Scineadora. In this model the transformation into an architectural static system emphasizes the expressive tension between the dancers.

*Performing Arts International*
1996, Vol. 1, Part 1, pp. 93–102
Reprints available directly from the publisher
Photocopying permitted by license only



# Signing in the Monsoon

Nigel Warrack

In travelling to the Nambikkai Foundation, South India, to work with young deaf adults, it has been possible to note and reflect upon their position and, by contrast, the equivalent in England. As a teacher and practitioner, I hoped to futher ideals promulgated by G. Sankala Pillai. Although an unexpected change of site occurred, newly organised workshops provided valuable insights. Tracking progress by use of a personal diary, the steps to achieving a dance/theatre dialogue with the handicapped from another culture can be seen.

KEY WORDS: Community dance, community theatre, south India, disability, deafness.

The Nambikkai Foundation is a rural community project in South India. The agricultural land is used for growing coconuts and limes, while the main complex houses dairy, poultry and goat units. All this provides resources for the area around Kanyakumari, the "Land's End" of India, where the Bay of Bengal meets the Indian Ocean. But the main function of Nambikkai is to provide education, training and employment opportunities for young deaf adults.

I first heard of the project in 1991 while working at another training institution, Rangha Prabhath Theatre Complex, 87 kilometres North, in Kerala. Rangha Prabhath was founded by the late G. Sankala Pillai for "the awareness and propagation of Gandhian ideals and values through drama among children and youth, particularly among the rural masses and weaker sections." There seemed to be potential for a connection between the two institutions.

I had been profoundly impressed with the vision of G. Sankala Pillai, his belief that no community was complete without a theatre along with a water supply, cultivable land and a place of worship. The creative energy of children, who "love to learn, but hate to be taught" is essential to the performances and the sense of co-operation between children and highly trained actors produces magical theatre. Much of my own work in the UK has involved disabled people, particularly with integrated groups such as Amici and Green Candle, and my intention

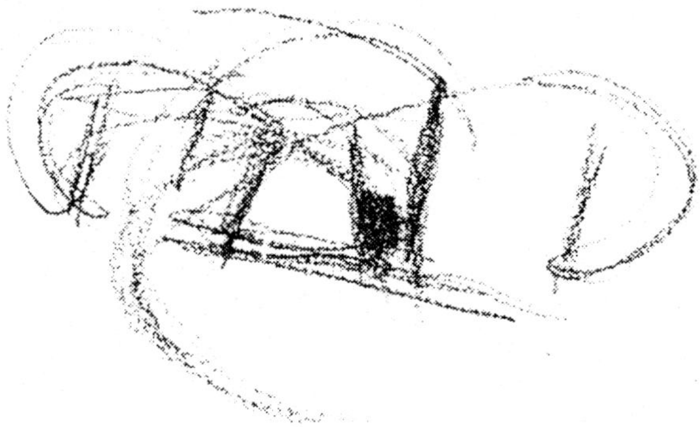

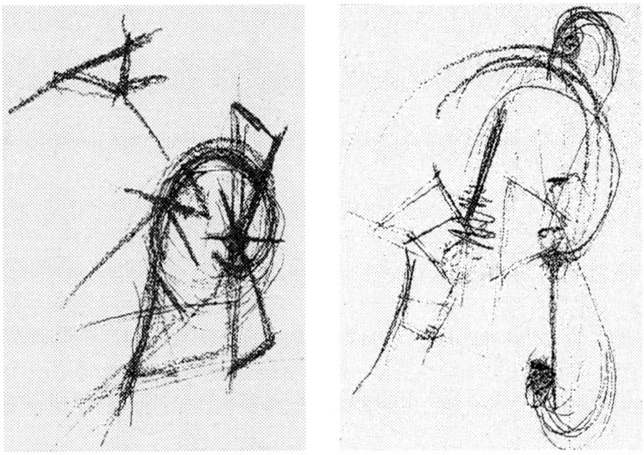

Drawings by Gallery/Michael Fiala.

showing a xylem (a channel through which water, absorbed by the root, is sucked up into the leaves; it is patterned like a spiral or a net which serves mainly to maintain its pipe-like shape). The supporting structure of the exhibition is derived from the xylem; its roots are assumed to be supporting channels of communication and the element of the globe comes from the idea of the cell of an organism.

On the other hand, the exhibition hall and the administrative area have contrasting geometric elements with visually and functionally opposite characters. The dance, developed out of these considerations, harmonised with the structural and functional requirements and delineated the spatial disposition of the architectural complex. Here, the practicability of this design process in reality could be seen, showing how architectural expression mediates contact between content and human intuition (*Peter Parolek*).

Exhibition Hall/Kurt Wilhelm. The movement of the male dancer is transformed into the volume of the exhibition hall; the forces are concentrated in a framework which is constructed through analysing the dancers.

Museum for Reptiles/Adolf Anzinger.

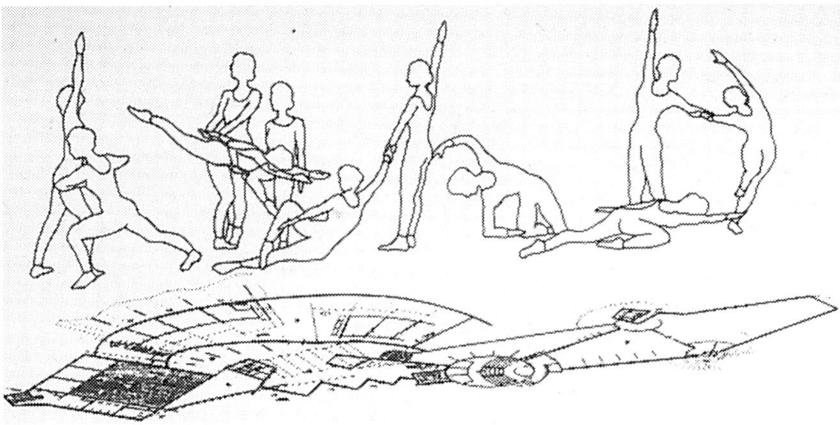

Drawings of dance by Renate Hammer. Projection by Peter Parolek.

## *Proportions*

In architecture, proportions are of great significance. Vitruvius (first century AD) wrote that the constructive members of a temple must have the same relation to the whole building as the members of the human body to the whole stature. It is interesting to note that, in the Middle Ages, Villar de Honnecourt worked with so-called 'homunculi' (human models), which were related to geometric forms and served as basic measures for the proportioning of cathedrals. The German music theorist, Hans Kayser, developed an example of Honnecourt's homunculi further to include the harmonic proportions formed in music – he saw harmonic architecture as a special kind of anthropometry, because it includes the human body (through its musical proportions) in the architecture (*Emmerich Simoncsics*).

For the purposes of these experiments, the principle was adopted that dance should be investigated as mere movement, entirely separate from music. In order to have the potential ability to carry out relevant movements for the artistic dance, a dancer needs physical training to achieve the necessary spectrum of movement. Through this the whole person is formed (physical condition, shape, and mental attitude).

When comparing dance with architecture, the first idea which comes to one's mind is their spatial kinship. Perceiving them, however, the spectator's position is different. During a performance, dance is structuring space in time (representing the third and fourth dimension), while the spectator does not change position. However, architecture can be perceived in time only when going along a facade or going through/ around a building.

A homunculus (13th century) and the geometric propotion applied for gothic cathedrals.

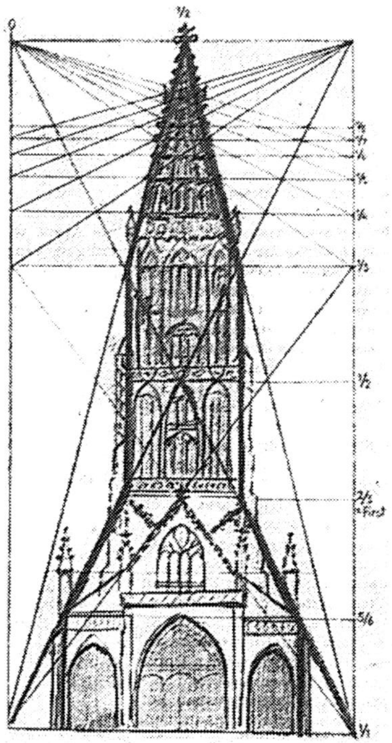

The geometric proportions of a gothic cathedral are derived from the harmonic proportions of the human body.

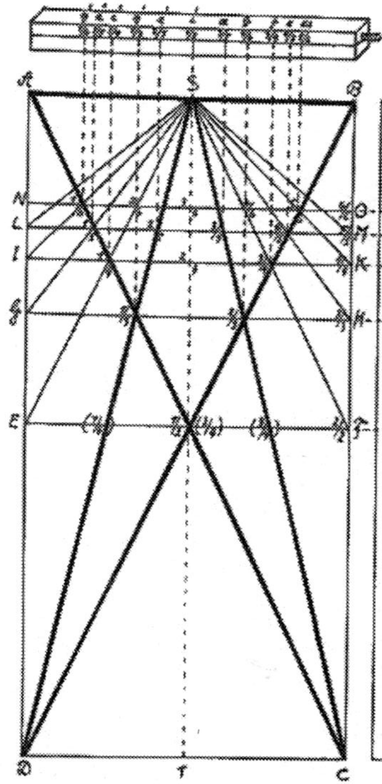

On the top and at the side by means of the monochord the harmonic proportions were determined. The proportions of the human body, music and architecture were meant to be in complete harmony.

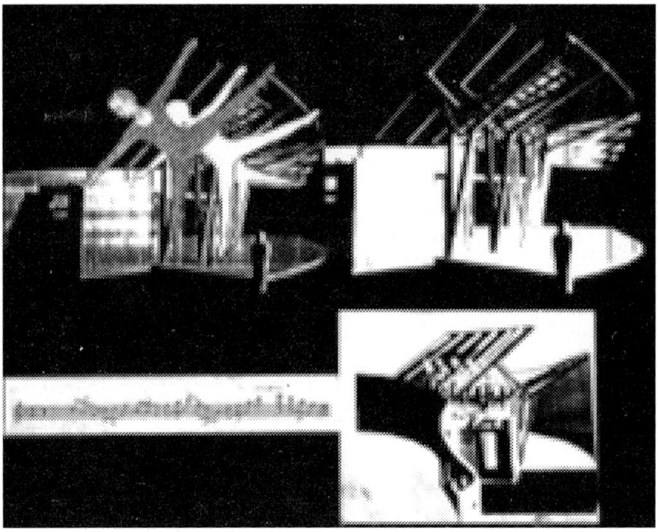

Gallery/Karl Gruber, Clarissa Leher, Werner Makus.

Movement is the basic element of dance. This dynamic feature, however, is only one possible form of expression in architecture. Perception is also a dynamic process. Dancers directly feel the transmission of force in their muscles; the sensation of movement in the muscles, joints and tendons is called the kinaesthetic sense. In architecture the quality of dynamic experience depends mainly on its spatial configuration, so that the architect has to learn to recreate physical sensation in the imagination in order to develop clear, static structural forms. While dance positions help us to analyze the static feature of architecture, movement in dance directs our attention to the dynamic character of spatial design. An intuitive perception of the three dimensional proportions of the human body takes place when dancers are observed in continuously changing positions.

One experiment concentrated on the space-structuring feature of dance in terms of a space, like an open air structure without a roof (space in space). The alternately changing movements describing force and space led to a spatial conception, a quasi-three-dimensional notation of the dance. Another had the intention to make a composition out of different geometric forms (objects in space). Through a sequence of movements were solids, the sculptural features of dance, to be transformed into a sculptural architecture. In these examples, dance was made in a highly abstract way, mainly considering the basic elements of architecture like space, force, or proportion. No expression of feeling, no gesture (body language) influenced the development of the purely tectonic elements derived from architectural design. It was evident that by the additional quality of movement, different characteristics and atmosphere can be achieved.

## *Alternation in Rhythm and Force*

The intensity of visual information, whether during the perception of a facade or in the course of a dance, can vary per time unit. Short and quick movements correspond to more information than extended and slow movements. Consequently, the architectural structure is more detailed in the first case than in the second, which rather initiates long extended walls without divisions into smaller elements. Positions expressing force and energy are contrary to movements carried out with lightness. Whereas in the first case, the architectural translation emphasizes tectonic feature, the second case initiates a light building mass with a floating or hovering character.

These parameters can induce a response from a composer, whose interpretation of movements can make us conscious of the fact that music and dance are mediums with similar characteristics in their capacity to express emotions. Architecture is an abstract art, but it can also evoke emotions. We can only comprehend the essence of this phenomenon by making comparisons with other art forms such as music, dance, or mime, which themselves are more directly linked with the human psyche (*Emmerich Simoncsics*).

## *Bibliography*

Daidalos: Danced Space, Conflicts of Modern Dance Theatre *Architektur Kunst Kultur* no. 44, p. 41.

Laban, Rudolf von: Choreutik, Grundlagen der Harmonielehre des Tanzes, Florian Notzel Verlag, 1966.

Laban, Rudolf von: Choreutiks: Grundlagen der Raumharmonielehre des Tanzes, Florian Notzel Verlag, 1991.

G Oberzaucher-Schuler: Eine Mitteleuropaische Bewegung der ersten Halfte des 20 Jahrhunderts, Florian Notzel Verlag 1992.

*Performing Arts International*
1996, Vol. 1, Part 1, pp. 53–66
Reprints available directly from the publisher
Photocopying permitted by license only



# Performance as Narrative – Notes Towards a Theory of the Performing Arts

Michael Bridger

The interest of narrative theorists in exploring links with performance is mirrored here by suggesting that useful insights into performance can result from considering ways in which it demonstrates characteristics of narrative. A range of perspectives are suggested which, with further elaboration, may form the basis of a comprehensive theory of the performing arts.

KEY WORDS: narratology, Marie Maclean, semiotics, bisociation, temporal processes.

"It was a dark and stormy night and Brigham Young and Brigham Old sat around the campfire. Tell us a story, old man! And this is the story he told:..."

(LeGuin, in Mitchell, 1980 p. 187)

So start the intertwined loops of Ursula LeGuin's virtuosic exploration-by-demonstration of 'the what, the how and the why of narrative', showing how layers of representation, meaning and time sequence can be nested within each other. In this article, my intention is to take consideration of the operation and significance of sequential processes of this type as a starting point, encompassing both narrative and performance, and to argue that this perspective of a combined narrative/performance theory is both fruitful and revealing. Not, of course, to claim that this, or any other mere theory can totally 'explain' either narrative or performance, both of which are characterised as much as anything by ambiguity, multiplicity and mystery, but to suggest that there are shared elements at a basic, structural level that satisfy aesthetic needs in many different media in both direct and symbolic ways.

The search for unified theories is characteristic of many fields of academic enquiry at the present time; examples range from the astrophysicist's quest for a theory encompassing all the principal forces observed in the physical universe, to the semiotician's search for a rationale underlying all sign systems. My own interest in a broad, generally applicable theory of the performing arts stems from a project in which I took the semiotic apparatus of Roland Barthes as the basis for an investigation into expressive and structural aspects of a group of works of electroacoustic music. More specifically, I modified the 'codes of signification' developed by him for the purpose of literary analysis in his book *S/Z* in order to create a common analytical and critical environment for a disparate group of works in the different, and somewhat problematic medium of electroacoustic music. It was, I think, a fruitful approach that at least had the merit of coping with contemporary material (most musical analysis, oddly, seems to focus on the pre-twentieth century tonal music that is so much easier to deal with). A problem I found, however, was that the methodology of semiotics seemed to have little to offer with respect to the analysis and codification of the temporal flow and articulation of music, which is clearly so vital a characteristic of the art. 'Signs' can be identified easily in electroacoustic music, as in any other style, but attention must also be given to the arrangement of these signs into dynamic structures in time, with complex hierarchies of relationship, repetition and differentiation, and semiotics offers little insight here.

Increasingly I have felt that narrative theory (narratology) must have something to offer both to the specific area of music theory and to performance theory generally, since narrative structures and processes display a complex linearity and sequentiality akin to that of performance. Many performing arts works are overtly narrative, so it could even be claimed with some justification that all performance is a sub-set of a very broadly defined genre of narrative, and it is clearly the case that dramatic works and films commonly employ the same narrative devices and strategies as literary forms such as the novel. Study of literary narrative has achieved a considerable degree of sophistication and credibility, so it seems both legitimate and promising to explore ways in which its concepts and methodology may be relevant to the performing arts. Related to the present investigation by a reversal of terms in the two titles, the book *Narrative as Performance* by Marie Maclean offers many insights that are readily transferable to the performing arts and much of what follows derives in one way or another from that stimulating book.

An important matter to consider initially, since I am speculating about a range of devices that underlie both performance and narrative, is the differing ways in which the two fields can relate to each other. Within the field of performance there are, of course, a number of immediate and direct links with narrative. Drama is essentially narrative in nature, dance commonly so and music occasionally. In other words, quite apart from the search for parallels between narrative and performance, narrative is often itself an element of performance in its own right – a fact that simplifies my task in some ways and complicates it in others. There is a sort of complementarity at work here, in which the more abstract forms of dance and music derive associative resonances from a sensed affinity with the semantically focussed narrative of literature and drama, which, in turn, benefits from links with the more aesthetically focussed, content-free rhetoric of dance and music. There is ample evidence that this transference between media has been consciously exploited by creative artists, but I would argue that there is often a subconscious element at work also.

There are, of course, both differences and similarities between the written narrative of the novel or poem and the performed narrative or quasi-narrative of drama, dance or music. My starting point is the hypothesis that by considering work in a very wide range of genres (that may initially seem unrelated) it will be possible to delineate a taxonomy of devices and processes applicable in some cases to both fields, in others more particularly to one, but nevertheless revealing a previously unexplored commonality. There are clearly many areas common to the performing arts that are susceptible to analysis and theoretical speculation; one could examine and compare stylistic trends as manifested in the various media, treatment of similar themes in different media; the roles of choreographer, playwright and composer or dancer, actor and musician; differing attitudes to the extension of style, technique and medium in avantgarde developments could be analysed, and the various fusions, combinations and interactions between media... In other words, one can envisage not just a single theory of performing arts, but rather a network of interlinking theories, which would reveal traits of the genre as a whole, as well as highlighting qualities of the individual arts by processes of contextualisation and differentiation.

My aim in this article, then, is not to attempt to define a coherent and comprehensive theory of the performing arts in a single step, but rather to reconnoitre the territory of such a theory, to identify some of its necessary features and to consider ways in which the performing arts are differentiated as well as some of their shared characteristics. My main topics are:

1 structure and temporal flow
2 solo performer, ensemble and chorus
3 performer/audience interaction
4 performer and text
5 paradox, ambiguity and metaphor

## *1 Structure and temporal flow*

As I implied earlier, the concept of structure, with its connotation of predetermined stasis, does not seem in itself a sufficient guiding principle for performing arts theory. This is especially the case with respect to the actual experience of performance. Processes involved in analysis, criticism and a work's initial creation are very much concerned with matters of structure, but this is because they operate in a frame of cerebral consideration outside the inexorable, real-time progression that so strongly characterises the performance event itself. Perhaps I am being over-restrictive in my interpretation of the term 'structure' but the inexorable, one-dimensional linearity of performance along the arrow of time seems to me so fundamental a characteristic, that it is fruitful to consider the applicability of alternative, time-based concepts such as sequence, articulation, repetition, progression, phasing, interlocking cycles of tension and resolution... All of these suggest organisation in time, whereby the virtual structure of a work becomes manifest in a temporal flow in which it is, as it were, reassembled as an evolving structural entity in the process of its reception by an audience. This process involves all sorts of abilities and expertises in an audience–familiarity with the idiom, memory, expectation, acuteness of perception, attentiveness etc... Additionally, there are specific short term aspects of temporal organisation in all performances – metre and rhythm in music, dance and dramatic dialogue (especially in verse plays), an actors's timing, the overall pace or tempo and its constant flux of minute modifications, sensitive to the particular circumstances, acoustics and audience feedback of each separate occasion.

Just as the visual arts control, formalise and distil the impact of the physical, three-dimensional world on our experience, so do the performing arts tame and manipulate the elusive and threatening fourth dimension of time. Adapting Marie Maclean's view of narrative to this field, performance is concerned with entropy, a marshalling of human resources against the seemingly random dispersion of our experience. Both the performing and visual arts do a good deal more than implied by this, I realise, but a quality of coming to terms with the elements of our existence surely lies at the heart of our fascination with all art.

Like the thread leading through the minotaur's maze, a narrative or a performance must in some way establish and maintain a strand of sequential, cumulative logic. This is not to say that its outcomes or on-going development should be totally predictable. The opposite, of course, is the case. As Ricoeur points out,

> "the story's conclusion is the pole of attraction of the entire development. There is no story if our attention is not moved along by a thousand contingencies. This is why a story has to be followed to its conclusion. So rather than being predictable, a conclusion must be acceptable."
>
> (Ricoeur, 1980 p. 170)

Predictability is not required, then, but rather a cohesive chain between beginning and end that negotiates a course through or around multiple possibilities. The image of a maze or labyrinth suggested above is a good analogy of this principle operating in its simplest way.

It is not the case, however, that narrative linearity corresponds exactly to the linearity of real or imagined events recalled or depicted, indeed this simple parallel is hardly ever found. Almost always earlier or future events are alluded to so that the reader/audience accumulates a penumbra of time consciousness around the on-going quasi-present of the main narrative. Occasionally, these conventions are deliberately upset, as in *Betrayal* by Pinter in which the course of an unhappy love affair is depicted in a series of episodes presented in reverse order. As is usually the case when conventions are disrupted, the effect of this is startling, disconcerting and extremely telling.

In a play, film or novel the guiding principle of linearity is clearly derived from our own everyday experience, in effect spinning out a fictionalised and stylised but nevertheless recognisable distillation of real life. In dance or music there is seldom so direct an external parallel to draw on, but in my view in both art forms an abstracted, nonspecific quasi-narrative underlies and underpins the onward processes of both composition and performance. More is involved in the creation of linear continuity than just a single strand, of course; perhaps it is best not to think of the narrative itself as being linear, but rather the reader/audience's apprehension and perception of it. Almost always the narrative itself will create and exploit deliberate discontinuities which, paradoxically, have the effect of reinforcing the reader/audience's sense of linearity as much as anything by imposing a stress upon it. Radical discontinuities are similarly apparent in most dance or music works, and I suspect that analysis would reveal a tendency towards comparable distributions if one were to plot periods of continuity and fractures of discontinuity in works in a variety of media. This is an unsupported hypothesis to be treated with due caution – an alternative outcome, if anyone were to undertake the necessary research, might well be to discover a range of distinct types of structure that run across boundaries of form and medium. Or, of course, nothing may be discovered at all that would be so conveniently supportive of my case – in the absence of detailed research only the idea of this new perspective can be put forward with confidence, but it is, I think a new and worthwhile one that for me, intrigued by the fascinating universality of fractals in the emergent science of chaos, has the added virture of meshing with a major paradigm shift of contemporary science. I am aware that to some extent I have employed a circular argument here, suggesting on the one hand an infrastructure akin to literary narrative in dance and music and on the other, an infrastructure of abstract principles beneath the surface narrative of the novel or drama. Chicken and egg reasoning I suppose, but the value of the chicken/egg quandary is precisely that either variant of the paradox provides a stimulating perspective and that neither is exculsively 'correct'.

Moving on to more general aspects related to the notion of a linear thread of continuity other factors demand consideration. For instance, by what mechanisms can the continuity be maintained, and by what sensibilities does the reader/audience judge that the continuity is valid rather than spurious or accidental? These two points are closely related, since, as we know from the tenets of Gestalt psychology, a characteristic of the human brain is to search for patterns and continuities amongst the miscellaneous sensory impressions that assail us at all times. So the reader/audience is predisposed to find patterns and continuities even where none exist – a fact that all creative artists exploit. The art of film, with its constant cutting between often totally unrelated images, would be impossible without this fundamental characteristic of human perception. Adopting the terminology of Roland Barthes in *S/Z*, a hermeneutic process is involved, and one further corollary of this is the necessity for plurality of possible outcomes, just as the correct route though a maze must necessarily be surrounded by false paths. Difficulty and complexity are essential characteristics.

In literary or dramatic narrative the main constant that sustains continuity is *character*, the collective of dramatis personae whose interaction provides the warp against the weft of external circumstance in the evolving plot. And, of course, it is easy to find paralles in music and dance performance, the simplest being the 'cast' of actual performers themselves, though their role in dance is different from that of the actor in drama, and in music even more fundamentally different. 'Character' has more subtle parallels than this however, in such features as the first/second subject differentiation of sonata form in music. The fact that a continuity of theme or character is expected by the reader/ audience is amply shown, on the basis of the exception proving the rule, by Hitchcock's film *Psycho*, in which the murder of the presumed central character in the film's first reel creates a disconcerting shock even with repeated viewing.

Having touched on aspects of linear continuity, my next category is its opposite – articulation. A range of discontinuities are employed in performance works, from the obvious large scale segmentation into acts or movements to short term processive divisions such as the phrase in music or dance. And, of course, the motor of most work in these media – rhythm. Just as in the case of linear processes, articulation at all these levels balances expectation and surprise, so that the short- and long-term rhythms of a piece of music are established, prolonged and fractured in fluid cycles that themselves blend predictability with unpredictability.

In most musical styles, rhythms and articulations are usually overlaid on metrical grids that remain stable over relatively long periods; it is interesting to speculate that by the same inverted logic that proclaims the chicken as the means for reproducing the egg, there is a sense in which the apparently substantive purpose of music, its thematic and tonal processes, may conversely be seen as the means by which temporal flow, embodied in metre, is made manifest and celebrated, rather than vice versa. Metre in music or poetry is in effect a regular cyclical activity, operating at a short time scale in a hierarchy of periodicities extending to complex verse forms in poetry and to the thematic recurrences of rondo form and recapitulations of sonata form in music. In all narrative or performance works, these elements of recurrence are essential characteristics of sequential structure, fulfilling important needs for stability, but a stability that consists of the refraction of established features through the evolving lens of complexity that accumulates as the work progresses. In other words, not a true stability at all, but an apparent stability that all the more tellingly betrays the impossibility of reversion. In narrative and performance, as in life itself, however many loops and cycles of reassurance we seek, the essential process is irreversible. The preoccupation of some composers with the palindrome appears at first sight to be an exception to this principle, but closer consideration reveals that the palindrome in music is in fact a powerful statement of the impossibility of mirror relationships in time, except in the more abstruse and speculative realms of particle physics or cosmology.

The actual moment or process of articulation between different sections or materials can have a wide range of characteristics, ranging from barely perceptible transition to abrupt disjunction. In most works a variety of types are employed, and this variety is, I suspect, essential for the continuing interest of the reader/audience. There are a number of other characteristics of narrative and performance that I will do no more than list now as a series of polar terms – characteristics to do with pace and flow such as acceleration/ deceleration, stasis/progression, fast/ slow, periodic/aperiodic, regular/irregular, smooth/abrasive. And so on. I suspect that examples of each of these could probably be found in most media and genres, and also that a number of interesting things could be said about them. However, I will not pursue this particular area now, but will turn from micro to macro issues, or, following my starting analogy, from the threads and knots by which a path through a maze is mapped to the maze itself. Like all analogies, this one is arrestingly revealing in some ways, but inept in others. Concentrating on the former quality for the moment, I am attracted by the contrived, balanced physicality of the maze and by its implicit challenge to persistence and ingenuity. It is a problem that encompasses its own solution in a web of artifice.

I have referred to the need for a work to encompass a variety of types of transition between its sections or materials, and, of course, there must be a similar variety between the tone or texture of the materials them selves, separated by those processes of transition. This leads to consideration of perhaps the most important insights arising from the perspective on narrative and performance processes that I have outlined. In the familiar narrative form of the joke, a final moment of fractured perception is prepared by a process that establishes the parameters and mechanism by which it will be achieved. In his book *The Act of Creation*, Arthur Koestler convincingly argued that the essential mechanism of the joke is what he termed "bisociation", in which two apparently unrelated ideas collide like two points on the matrix of a graph. "Do you get the joke?" we enquire, and it is interesting to look within our own experience at what is implied by that bald word 'get'.

It would be pressing the point too far to imply that one can 'get' the first movement of Beethoven's *Eroica Symphony*, but in common with most works, that movement has a central climactic, even cataclysmic, moment not unlike the perception shift of a joke, though it is not, of course, funny. The overall musical flow of the movement pivots around that single moment, preparing, then resolving. A vital part of the composer's art is to judge the proportions of these processes. It is often tempting in personal listening to cut seemingly extraneous sections and to select just the highpoints, but this usually proves to be a most unsatisfactory experience. The need for some prolongation in the preparatory phase is well understood by the talented joke teller.

A range of models for the overall trajectory of a work of narrative or performance can be identified. The joke and 'whodunit', for instance, depend on a single sweep towards punchline or denouement, though there are commonly many loops and diversions on the way, like the false routes of a maze, which I have suggested play as important a role as the one correct route. The symphonic movement or play usually displays a more diffuse and complex flow profile. A very common model moves, like the joke, towards a climactic moment, but follows this with a shorter process that can best be described as a winding down or a re-establishment of stability. At the end of many works there is an indefinable sense that the work is relinquishing its hold on us and returning us to our normal world and life. Occasionally, with a very few of the highest achievements of performing arts, one almost has the alarming sense that this transition is taking place in reverse, that it is the work of art that is the higher reality and one's own life the illusion. Such is the power of art!

## 2 *Solo Performer, Ensemble and Chorus*

In much of the West European tradition, the role of performer is differentiated from that of the original creator of a performance work (for my present purpose I am excluding improvisatory performance or cases where the creator and performer are one and the same). The mingled elements of creative virtuoso and faithful, transparent intermediary lend a piquant ambiguity to the performer's role, whether actor, dancer or musician, and any systematic study of performing arts would need to tease apart a very complex web indeed in this area. For the moment I will attempt to unravel just one small corner, and that is the apparently logistic-based continuum of solo – ensemble – chorus that is evident in all performance media. As I have implied, a dual viewpoint is required – in one direction seeking areas of commonality between the three arts, in the other areas of differentiation.

Having noted the obvious similarity that the solo in dance or music is paralleled by the soliloquy in drama, a difference becomes immediately apparent, which derives from the most basic characteristics of the media. What I have in mind is to some extent a matter of definition, I realise, but it is that whereas in both dance and music several strands of equivalent, even unrelated activity can coexist, in drama, if one regards the thread of language as its principal sustaining factor, one finds a singularity rather than a multiplicity. Nevertheless, just as in dance and music, the essence of the solo is defined in part by what is missing, so that a common characteristic of the soliloquy is the internalised dialogue. I am, of course, referring to solo sections within ensemble works – the self-contained solo or the one-person show raise somewhat different issues.

The differentiation between the other two categories I have identified – the ensemble and the chorus, needs some elaboration. I take 'ensemble' to mean a smallish group of performers who work together, but with a strong element of dialogue and soloistic differentiation between them. In contrast, the chorus in performance is characterised by the involvement of a fairly large group, who work together, and whose individuality is to some extent sacrificed in order to accentuate their corporate cohesion and unanimity, which is often very precisely drilled. A characteristic evident in the chorus of Greek drama is the way in which it is externalised from the principal dynamic of the dramatic action, typically weighing human destiny in a wavering balance between destruction and reconciliation. In this dynamic, the deeper-level structural function of the chorus often differs from its apparent narrative role, by the paradox and artifice of dramatic construction.

The chorus/solo dichotomy seems evident in very equivalent ways in the various performance media. In the case of the conjunction of soloist and orchestral tutti in the musical form of the concerto, it is interesting to reflect on the intriguing and revealing philological ambiguity that surrounds the term 'concerto', which may imply either struggle or collaboration, depending on which derivation from Latin is preferred.

Oddly, both have appropriate connotations, since the tension between opposition and complementarity provides an element of paradox and ambiguity in performance that has much expressive potential.

Factors to be borne in mind in examining the solo – ensemble – chorus continuum of performance, then, include notions of opposition, collaboration and textural density. Clearly, in dance and music, but also, less conspicuously, in drama, a further element is that of virtuosity. The importance of virtuosity in the soloist's role is readily apparent, but the chorus has a virtuosity of its own, typically centred on synchronisation. This clearly involves some subjugation of individual display, but an important effect of this is to enhance its role as a foil to the different, highly individualistic virtuosity of the soloist.

## 3 *Performer – Audience Interactions*

Extending the notion of relationships in performance outwards towards the audience, a different interaction is involved, one that is by no means as passive as it sometimes seems. As Marie Maclean says:

> "Through a narrative text 'I' meet 'you' in a struggle which may be cooperative or may be combative, a struggle for knowledge, for power, for pleasure, for possession. The meeting is manifest in the course of the narrative performance, in which the performer undertakes to control the audience by visual or auditory signs while they the partners in the act, use their power as hearers or watchers to dictate the terms of that control. If you tell me a story, I can refuse to listen, but if I become a listener, even a forced listener, I can also always remind you that the story can only mean what my mind allows it to mean. I, too, am constantly performing... Narrative performance thus involves a network of intimate relationships between its participants which, like all such relationships, is at once a cooperation and a contest, an exercise in harmony and a mutual display of power. It is both act and interaction, and implies a contract, a recognition of obligation and expectation, thus acknowledging the rules which govern the interplay."
>
> (Maclean, p. xii)

The narrator/narratee, or performer/audience relationship is clearly one of considerable subtlety and complexity, and an important factor of that complexity is the sense of community between audience members. When one considers the poor acoustics, bad sight lines and general discomfort that are too often the audience's lot, the survival and continuing success of live performance at the present time, against the competition of television, video and hi-fi in the comfort of the home, must reveal that something of fundamental importance is involved in this need for community of experience. All the inconveniences I have mentioned very definitely interfere with the audience's reception of a performance and can be seen as a form of noise in the communication system. However, borrowing another idea from Marie Maclean, this noise may not be entirely a negative factor, functioning like the grit in the oyster that produces the pearl. An important element in the complex psychology of the audience is, I think, that although its individual members share the same experience, they do not share it identically, but each from a unique viewpoint and with different apprehensions and misapprehensions. As in so many aspects of life, unity and diversity coexist in a Yin-Yang-like clasp.

But there is, of course, a much more straightforward rationale for the attraction of mass attendance at live performances. Human beings are social creatures, needing each others' company in most aspects of everyday life, and having occasional needs for larger gatherings, whether family, tribal or religious in nature. Performing arts events, looked at in this way, can be seen to share many characteristics of other similar events; one can see in them traces of quasi-religious ceremonial, of gladiatorial display and daring, of a shared response to mythologised, transformed or quasi-realistic representations of life events or historical circumstances.

## 4 *Performer and text*

The two parties to the contract of performance, the performer and the audience, must be seen not only in their relationship to each other but also in their relationship to the text or the work. It is here, not surprisingly, that narrative theory provides some particularly pertinent notions. In the following slightly modified quotations from Marie Maclean, as in other places in this article, I have made only those changes of terminology necessary to move from the telling of tales to the performing of works:

"The performer has the privilege of speaking at the same time in the omniscient voice of the authoritative medium through which the work imposes itself, but also in his/her own physical voice and body, the 'I' present and performing, and the 'I' whose context and territory include the work itself and the licence to perform it. One must remember that through the performance is communicated not just the substance of the work but also the conventions of the medium, of stylistic tradition, of ideology, of the whole social context. So performers are at once external to the work and internal to it. Linked with the multiple voices of the performance is a multiplication of the focuses of reception in the audience, who perceive the work's message, while also receiving and judging the variations in style, in technique, in presentational savoir-faire, provided by the performance. The performance, in other words, functions as metatext to the audience, whose attention moves, in a form of internal dialogue, between the work and the performance. Their judgment is an active one, constantly providing feedback on the demands of variability versus invariability, on the delight of surprise as opposed to the pleasures of the formulaic. Their expectations of outcome may be fulfilled or unfulfilled (Brémond, 1973), to their approbation or chagrin... There exists within the same people an audience of the énonciation as well as an audience of the énonce."

(Maclean, 1988 pp. 7–9)

There are, of course, many types of relationship between performer and text. I have referred above to the role of performer as 'transparent intermediary', but this is only one end of a complex continuum and it is in any case an unattainable concept. Consider, for instance, the role of Hamlet and the widely diverse revelations of character, mood and motivation that are brought to the same fabric of words. It is interesting at this point to look sideways at other performance media to see whether the same range is evident in performances of the same text; it seems to me that individuality and differentiation is less marked, though nevertheless an important factor. Performances of, say, the *Hammerklavier Sonata* vary in many details and in overall conception and can be strongly characterised in different ways, but within a drama.

Especially in music, the concept of authenticity is currently and important aspect of performance, with a renewed attention to the rehabilitation of accurate performance practice that started with medieval music now reaching to the last years of the nineteenth century. Similar initiatives are evident in dance and drama – consider the construction of a replica of Shakespeare's Globe Theatre and the performance of an accurate reconstruction of Nijinsky's choreography for the *Rite of Spring*. Those who find the concept of postmodernism helpful in diagnosing late twentieth century culture will find in all of these scholarly, museum-like preservations of the past a reflection of the loss of confidence and essential nostalgia that characterises, or some would say afflicts, our contemporary sensibilities.

At the other end of the spectrum are the many instances when performers or performances impose deliberate distortions or dislocations on a text or on performance conventions. Modern dress versions of Shakespeare, Pina Bausch's choreography for The *Rite of Spring*, synthesizer versions of Bach keyboard music – in all such cases, the intention is far removed from the relatively anonymous, passive aesthetics of authenticity. Instead, a new revelation, even an affront, is intended – another dimension of postmodernism.

Especially in music, the term 'interpretation' is used to describe the performers's role, and this conveys something of the radical nature of transformation from the notation of a text into live performance, just as an interpreter translates between two discrete languages. Even in so-called 'authenticity', much guesswork and invention is involved, often with only a veneer of solid, definite historical evidence. And in the actual notations available for the performing arts much is omitted – the inflections of speech, actual stage movement in drama, gesture, fine details of musical dynamics and articulation... Approximation rather than predetermined precision is the main characteristic, and it is for this reason that the role of performer is so creative and diverse.

## *5 Paradox, Ambiguity and Metaphor*

An intrinsic characteristic of all art is that its communication is to an extent indirect, refracted through layers of suggestion, even deliberate obscurity, so as to impede the emergence of a single irrefutable meaning or response. The endless, unresolved debates about the meaning of music, or even about whether it is possible for music to have meaning at all, are a clear example of this. It seems to me that it is fundamental to any art medium that it should attempt things beyond its capabilities. Thus paintings and sculptures so often strive to represent movement, novels and poems to evoke visual scenes. In the case of music, its quasi-linguistic qualities are an example of this same impossible quest; music has its own grammar, with rules of succession and agreement, it has elements that resemble syntactical constructions, but all without the semantic element of relatively precise communication that is the raison d'être for language. Dance, too, is an articulated succession in time that seems to obey quasi-linguistic laws. For the audience, whose perception and thought processes are formed and conditioned by the apparatus of linguistic communication, what remains is the sense of language's grid of inexorable logic, of structured, meaningful sequence, but functioning as a carried not for semantic precision, but, in its absence (all the more powerfully) for emotional suggestion.

Drama, of course, employs language directly, so my preceding remarks need recasting if they are to be relevant to this medium. The point I was making was that artistic creation and communication sets up, at a basic level, a network of difficulties, against which ingenuity, imagination and technique struggle to break through to expressivity. Furthermore, I suggested that amongst these difficulties is one that cannot be overcome, an impossibility, and it is this insurmountable barrier that is the central fulcrum of a process of metaphor and paradox. At first sight, drama perhaps seems to lack this quality, but it is, I think, present in a particularly powerful way. The artifice of dramatic representation, after all, embodies the uniquely significant impossibility, that a person can become someone else. Whilst seeming to its audience a simulacrum of a real life, the essential impossibility of what we see on stage, and our willing complicity in its fiction, is as vital to drama as is music's impossible quest for language to that art.

The first of Barthes' codes of signification elaborated in *S/Z* the hermeneutic code, is relevant in this context, since it too is concerned with the notion of the importance of a quest for meaning in response to art, of a struggle to decipher, to quarry significance from an often opaque and intractable medium. The close correspondence between processes of literary narrative and performance is made apparent by their shared employment of a teasing mix of concealment and revelation. In literary and dramatic works of arts, the operation and interaction of these elements is especially crucial (the artifice surrounding the murderer's identity in the whodunit is just one, rather extreme example of this), but in dance and music also there is an equivalent tightly controlled and deliberately convoluted flow of a work's material.

The sustainment of mystery and ambiguity, then, is central to all art involving performance or narrative and the manipulation of audience response correspondingly must be a preoccupation of the work's creator, who must maintain an audience's interest in (and faith in the value of) what is revealed as assiduously as that revelation is delayed and confounded by deliberate false trails and digressions. If an essential factor in narrative is this interlinking with time in a sequential but convoluted revelation, a further temporal quality of narrative is its sense of a pre-existing reality that is recounted, a referent beyond the immediate narrative that is both revealed and apotheosised in that revelation. Performance similarly carries an aura of idealised reference, often identifiable with some precision in theatre or dance work, less so in music. In performances a real, imaginged or symbolically represented past becomes invested with iconic significance. Time, then, is in a very real sense not only the medium but also, with all its mysteries, the subject of both narrative and performance.

In the course of this article I have touched on a number of elements that must be accommodated within the conjectured narrative/performance theory I am concerned with, but acknowledgement of the predominant role of time processes has underlain all the various aspects of performance that I have considered. There is particular scope in this area, I am sure, for development of a fully articulated taxonomy of strategies and devices that will be applicable to drama, dance and music and that will give new insights by relating them to the fundamental human preoccupation with the telling of tales.

## Bibliography

Barthes, R. (1975) *S/Z*. (tr Richard Miller). London: Cape
Brémond, Claude (1973) *Logique du Récit*, Paris: Seuil
Koestler, A. (1976) *The Act of Creation*. London: Hutchinson
LeGuin, Ursula (1980) in *On Narrative* ed. Mitchell.W. Chicago: University of Chicago Press
Maclean, Marie (1988) *Narrative as Performance; the Baudelairean Experiment*, London: Routledge
Ricoeur, P. (1980) Narrative Time in *On Narrative* ed. Mitchell, W. Chicago: University of Chicago Press

*Performing Arts International*
1996, Vol. 1, Part 1, pp. 67–71
Reprints available directly from the publisher
Photocopying permitted by license only



# Currents of Influence in Japanese Theatre

Toshihiko Norizuki

An examination of the fusion of Western conventions and Japanese classical traditions during 19th and 20th centuries that led to today's Japanese theatre. Key features of Western and Japanese drama merged to form today's hybrid styles. In the context of underlying social and political changes, Japanese theatre has in effect now reached the point where the very act of communication is questioned.

KEY WORDS: Japanese theatre, Shingeki, Kabuki, New Wave theatre.

Today's Japanese theatre is almost the same as western theatre with respect to staging style. While using the Japanese language, it uses westernized drama, westernized costume and western theatrical technique: nevertheless there are still many differences between western and Japanese theatre. For example, the way of thinking about dramatic conflict differs. Most of today's Japanese theatre, particularly young people's theatre written by young Japanese dramatists, seeks the dramatic conflict in a common illusion, and not in an individual element. Perhaps, this is due to the totalitarianism which is still prevalent in the educational world in Japan. Traditionally, western theatre sets up conflict between two or more people, or a person versus society or other groups. So, when western people see today's Japanese theatre, they may not see any dramatic conflict in the drama. There isn't an individual's conflict in the drama.

On the other hand, there are many differences between presenting a play in western style or in Japanese style on stage. For example, when an actor enters from stage-left, audiences understand that the actor comes from inside a building. When an actor enters from the right hand side the actor comes from outside a building. Or when a story reaches its climax a dance scene suddenly begins (the same as in Indian films). These are unspoken conventions which came from Noh and Kabuki theatre. Such traditional elements are still lingering in today's theatre without young theatregoers' awareness.

In this article, I outline the history of modern theatre in Japan, showing the interaction of classical Japanese and western influences.

## *A short history of the beginnings of Shingeki*

In 1868, the seclusion policy by the Tokugawa government was abolished; Japan opened its doors to the U.S.A. and other countries and the Meiji period began. During the Meiji restoration numerous aspects of western civilization, from daily necessities to political structure, were imported. People wanted many things from the modern west at that time.

Kabuki was alive during the Meiji restoration and an attempt was made to arrange western drama in Kabuki-style. But Noh wasn't alive at that time because its patrons, the Tokugawa Shogunate and the Samurai (warrior) class, had their downfall. Noh was regenerated by a new noble class (the former lower warrior class) of the Meiji period after 1880. But they weren't motivated to adapt western theatre to their stage – they kept to their classical style only.

In the case of Kabuki, to conform with the new times, western drama was modified for the Kabuki stage. For example, an English play, *Money* (1840, performed at the Theatre Royal Haymarket, London) by Edward Bulwer-Lytton (1803–1873), was modified in the Kabuki style and the result was *"Ningen banji kane no yononaka* (*All is money*)*"* (1879) by Kawatake Moku'ami (1816–1893) who was a well-known Kabuki dramatist at that time. All elements of the original, places, names of characters and so on, were changed to Japanese equivalents and the play was performed in the Kabuki style accompanied by Shami-sen music (a three'stringed banjo'like instrument). The actors wore the Meiji-style Kimono and their makeup was natural.

During the Meiji period (19th century), some new styles of theatre appeared, such as womens' theatre (Onna Shibai), students' theatre (Shosei Shibai), and others. These new styles were highly popular among audiences. They retained Kabuki-style staging, Shami-sen music and other elements. The students' theatre attempted to modify many western dramas, including Shakespeare, for their theatre called Shingeki (literally, new theatre). This was the first example of the term Shingeki – today, we call it Shin-pa, literally "new school", because we evaluate it as one of the new schools of Kabuki. An original theatrical world equal to Kabuki was developed by the beginning of the 20th century.

However, at that time, the motion picture was imported from the U.S.A., and spread all over Japan very rapidly. Soon, nearly half of the theatres were occupied by movies, so that Kabuki and Shingeki (Shin-pa) had to reduce the number of performances and their energy. They gradually stopped creating new plays, relying instead on classical ones.

## *The time of anti-Kabuki*

At the beginning of the 20th century, another approach of Shingeki performance to western drama appeared. Rather than modifying plays, they were translated. Examples included Shakespeare, Henrik Ibsen (1828–1906, the Norwegian playwright regarded as the father of 20th century drama) and other plays chosen at random and unrelated to each other. Unfortunately, the differing styles and levels of directing between, for example, Shakespeare and Ibsen were ignored. Productions avoided Kabuki technique and applied western costume and makeup. Western acting skills were adopted in order to imitate western methods of movement and hand gesture, and the behaviour and expression of western people. The slogan was "anti-Kabuki", perhaps this point of view is still alive in Japan today – as I have heard students say, "Kabuki is not drama, it is only musical comedy". This attitude comes from "anti-Kabuki", illustrating the break with Kabuki and traditional theatre technique. Another factor was the view that Shingeki was not commercialized theatre like Kabuki.

Shingeki began without professional acting, directing, or dramatization – it was the amateur theatre of that time. But the intelligentsia welcomed the amateurish Shingeki because it was a simple way to see the western mind displayed in drama – easier to comprehend than by reading a book from the west.

Shingeki developed gradually. Unwittingly, the exponents acquired and combined theatrical techniques from western theatre and classical Japanese theatre. For example, during a monologue they used western electric lighting, but in a very Japanese style, focussing a spot light on the actor speaking the monologue and fading out all other lights. This lighting direction suggests that the actor speaks with an inner voice – a similar effect to the close-up in films. This method of lighting reflects Kabuki lighting of the Edo period which did the same but with candles ("Sashi-dashi" is a Kabuki term which means hold out the candle or "Tsura-akari" which means lighting to the face). Perhaps they learnt this method from western films and didn't know that it also came from the Kabuki.

After 1920, original Japanese plays influenced by western plays were written. In the 1930s, the Shingeki split into two strands of opinion: one of them was "art for art's sake" and the other one was "socialism". Their audiences were divided–the "art for art's sake" type of Shingeki had an audience which was still the intelligentsia and young people who longed for art, while the "socialism" type of Shingeki found a new audience in workers striving for social reformation.

The second world war had a major impact. During it, many productions were prohibited from being performed except some that followed the national policy. While after the second world war, both political and social conditions changed radically in Japan, in effect controlled by the U.S.A. after the two countries signed a security treaty. The young generation opposed the security treaty because they thought that this was a disadvantageous treaty for the Japanese and they were concerned that they would lose their Japanese identity. In those days, the U.S.A. was at war against the U.S.S.R. in Korea and in Vietnam and its military bases were in Japan. The 1960s theatre evolved in this political context.

## *The time of anti-Shingeki*

The theatre of the 1960s was named Angura-engeki ("underground theatre"). Angura was avant-grade and against Shingeki, so some critics call it post-Shingeki. Angura, or post-Shingeki, theatres and dramas caused great waves of influence on present theatre.

Angura played original drama in non-theatrical venues. They couldn't use well-known theatres because they didn't have enough money to rent them. Instead they presented performances in small spaces which were low rent or free. For example, a public square or the grounds of a shrine where they pitched a big tent, a storehouse, a tea room, etc. Similar places are found in London or in New York, called 'fringe' in England and 'off-broadway' in the U.S.A. So, their starting point was 'an empty space'. They used only second-hand old clothes which were collected from their families or friends for their stage costumes and didn't have sophisticated theatrical equipment, but they nevertheless had some original ideas about drama and theatre.

Techniques from the Noh and the Kabuki were consciously adapted, and for this reason they didn't have any prejudice against the Japanese traditional theatre in the way that Shingeki had; instead they made original plays which took the best of the traditional theatre. For example, they called one of their acting areas "Hana-michi" (literally, flower path, one of the Kabuki terms). It was a stage mechanism in Kabuki but their version of 'Hana-michi' wasn't a stage mechanism, meaning instead that they performed in the middle of the audience who were sitting on the floor. Or sometimes they used lines of classical Kabuki scripts in their plays and they often used the Noh dramaturgical device of changing the time either back to the past or to the future in a flash. Also they often used popular music, particularly Enka, Japanese popular ballads, popular songs and so on. This kind of music works to create a romantic atmosphere and to express a character's appearance and emotion. Perhaps this element also came from the Kabuki.

There were many exponents of Angura theatre in the 1960s who, although they didn't have many common features, proclaimed that theirs was the only true contemporary theatre, and that the Shingeki had already died because it had no ideas to express or to represent to the age, which included the political and social conditions of the '60s. Cross-fertilisation occurred however – staging styles of Angura influenced the Shingeki and the Kabuki, who began to use small spaces for their performances, while the Shingeki began to perform the Angura plays because they were so vivid as contemporary theatre.

## *The new wave theatre*

After the 1970s, a younger generation's theatre appeared; its staging was similar to the Angura theatre, performed in non-theatrical places and using popular music. But it was indifferent to the political situation, especially the security treaty between the U.S.A. and Japan. The younger generation already knew that nothing had changed as a result of the security treaty through the opposition movement of the '60s. It was a non-political generation, interested in its own problems – love, sex, the entrance examination of university and success in life. And they were isolated because all other people of the same generation were rivals for the ordeal of the entrance examinations. It was difficult to communicate with each other.

So the younger generation's theatre scooped this situation up and expressed it on stage. For example, the 1970s dramatists wrote of types of situations found in comic books, and they used many characters from comic books as well. Comic books were like best friends to the isolated younger generation – it seemed as though perhaps they could only communicate through comic book characters. The main subject of this drama was isolation from others and society – alienation. But their stages are cheerful and bright in the illusion of their hope. Any dreams they have will be realized on stage.

*Performing Arts International*
1996, Vol. 1, Part 1, pp. 73–77
Reprints available directly from the publisher
Photocopying permitted by license only



# Text, Context, Dance

Shobana Jeyasingh

Significant inequalities exist between the media representations of East and West culture, glaringly obvious with the advent of global communications. Specifically, in dance there are inequalities. And yet in practice, as a choreographer, it has been possible to bring together diverse cultural and geographical heritages. One early piece *The Making of Maps* successfully merged Indian dance with East and West found sounds and music. Even in a traditional Indian dance language, such as Bharatha Natyam, it has been possible to fuse Western music with the dictates of the dance. It is possible, as a Londoner, to meld together Indian heritage and Urban up-to-date concerns.

KEY WORDS: Dance, Indian dance, Bharatha Natyam, multiculturalism, East and West culture.

A recent lavish series on dance which was broadcast on British television interviewed a Bharatha Natyam dancer from Madras. Although the Madras traffic could be heard in the no doubt carefully prepared soundtrack, the camera cut to a rice field (nowhere near the offices and light industry of Madras) and came to rest happily on a water droplet falling gently from a fresh green stem. The background talk was Gods, Goddesses and spiritual pursuits – it had to be since the producers (not Indian) had already decided that Bharatha Natyam was to come under the heading of 'religion'. This despite the fact that it is performed in theatres not only in Madras but in major cities all over the world. These interesting complications of history, such as the fact that the Bharatha Natyam dancer so exotically designed into the frame spoke impeccable English, and that she probably had a degree in Computer Science, or that this 'pure and religious' art form thrived within a traffic-infested city were ignored. The function of 'the orient', so beloved in the West as a place of simple spiritual certainties, exotic maidens and colourful rituals, had to be reinforced once again to the viewers in their sitting rooms in London and New York. There is simply no room for the realities of carbon monoxide in a romance.

This stifling, unhistoric and ultimately life-denying concept of the East is not an isolated example. It is very much a symptom of a more general, unequal power relationship between East and West. The West is the external anthropologist with the resources to observe, research and 'explain' all those exotic cultures. It is a kind of colonization through categorising and as such is an exercise of power. There are not many programmes on Indian television about the natives of Tunbridge Wells but, numerous ones on British television on selected aspects of Indian life. It is Columbus who discovers America simply because his decendants have the power to control the telling of the story.

The evaluation and understanding of South Asian dance is often hampered by assumptions which are the by-products of this inequality. As a choreographer, my work is described as an East/West collaboration because I work with composers who are not Indian. The fact that we both share a common British culture is never highlighted. The assumption is that 'East', myself, must be a simple unchanging essence which stands for tradition and The Past, and 'West', the collaborator, represents change, modernity and dynamism. It is also symptomatic that when the 'unchanging East' is influenced by 'the West' it is always news. However when 'the West' is influenced by Indian musical structure (as Phillip Glass was) or by African art (as Picasso was) or South Asian dance (as Martha Graham was) it is hardly something to comment on or acknowledge. One very rarely hears of a West/East collaboration.

This polarised and inaccurate view of other cultures is just about understandable in previous centuries, when the geographical borders defined cultural identities and there was little chance to test one's ideas of other cultures in practice. The 'India' that is the backdrop to a ballet like *La Bayadere* with its kitsch orientalism and characters who could be anywhere from Cairo to Tokyo, is irksome to an Indian but understandable as a product of its time. It is less understandable to subscribe to such a landscape in the late twentieth century, when the post colonial migration of peoples all over the globe have resulted in the blurring of a simple East/West divide, especially as a cultural reference point. Migrations have occurred not only from the so-called third world to Europe and America but from Britain to Australia and within Europe itself. Even within political and national boundaries people have moved from rural, provincial settings to ever increasing urban ones. One of the features of this century is that very few people live their lives like their parents and grandparents lived. The revolution in communication has further meant that people have set up global communities of common interest – whether it be of high finance or of personal predilection.

One such global community is the Indian diaspora which stretches from Fiji to Canada. The culture pertinent to it is characterised by a balancing act between old and new; memory and present day reality.

The members of this diverse community live their life in a cultural roundabout where many histories meet. A continuous stream of new arrivals at differing states of acclimatisation makes sure that nothing stays fixed for long. The choice as to what road to take – preservation, innovation, community, mainstream, integration, seclusion or a mix of any of these are all possibilities – remains a purely individual one. The culture of the roundabout is historically new terrain and its landscape is just beginning to be chartered. One can glimpse it in the multiple reference points in a Rushdie novel and in the tragi-comedy of films like *Bhaji on the Beach*. The sculptures of Anish Kapoor, in their very transcendence of cultural specifies, declare themselves as a product of the roundabout as does the rap music of Apache Indian. The culture of the diaspora challenges inherited ideas of home as something defined by geography. More and more, 'home' is becoming a radical and dynamic invention that is more about the future than the past – the sum of new journeys rather than the station one has left behind.

It was my personal experience of homemaking that prompted me to choreograph one of my earliest dance works *The Making of Maps*. As the title suggests it was not the end product but the process that I was interested in. I wanted to challenge the total inaccuracy of the 'home ground' that was prescribed for me. I had spent my childhood and youth in four different countries and my artistic heritage was therefore diverse. The one thing that connected all the strands was the fact that I had always been a city dweller and felt at home in that particular pacing of life. The score for *Making of Maps* was made up of found sounds of city – the hum of conversation, traffic, railway stations as well as domestic sounds. This was composed by Alistair McDonald from Birmingham. Woven into the composition was a fragment of a carnatic music ensemble rehearsing. The ensemble in fact was put together by the composer Ramamani who lives in Bangalore, South India. The music she composed for the piece emerged as quote within the overall composition, and it had the accessible entrance that the age of electronics has given to all types of music. At the press of a button anywhere in the world, either through the radio or a music centre, one could create the intimacy of a performance while acknowledging its separateness from one's surroundings. In fact, the way one could twiddle the knob of a radio and travel through diverse soundscapes was a metaphor that was used in the creation of the music. The manner in which McDonald's music related to Ramamani's was important. The Indian music was not designed to be a dislocated or nostalgic episode and neither were the city sounds intended to be a negative or threatening force.

In terms of dance language our starting point was Bharatha Natyam – that was a given like race or gender. Just as the sounds that accompanied the dance moved from personal to formal, the dance too was influenced by purely personal concerns of vocabulary and composition. Traditionally Bharatha Natyam has a close affinity to the floor and to the pull of gravity. The body 'sits' in a demi-plier position and energizes itself through rhythmic contact between feet and ground. I wanted to push this affinity to its logical conclusion and had my dancers travel through demi-plier, slide onto the floor and roll over. This embrace of the floor signalled a loosening of the prescribed 'poise' that is part of the etiquette of the dance. This in turn made it easier to question, however tentatively, the self-sufficiency of the Bharatha Natyam body conditioned by its history as a solo art form. It was not just a matter of introducing touch and with it the implication of human emotion and relationships, but also the acceptance of physical dependency and trust. Dancers had to be helped off the floor if the dance were to continue, pushed to the ground or helped to jump further. It was important that these interactions were casual as well as stylized.

In the music McDonald had engineered an extremely organic entry into the Indian music and the exit out of it. I did not want the dance language to change dramatically when we danced to Ramamani's music – that would have made that section seem either like a romantic and exotic interlude or made it more 'real' than the city sounds. Therefore whatever discoveries we had made in terms of vocabulary and composition were common to both. The changes in the dance, in terms of density of structure or speed, were in response to the differing musical structures rather than any perceived cultural changes. Carnatic music composition by its very nature dictates the ebb and flow of movement. McDonald's music by contrast afforded the movement the freedom to breathe and find its own pacing.

It seemed fitting that *Making of Maps* should end with the sounds of the city. Usually these are used to evoke fragmentation, dissonance and disturbance. However, personally I have no negative feelings about living in London and neither do I feel that the sense of order, which I find appealing in Bharatha Natyam, undermined by it. The music takes the dance into a time and place in the city when the sounds are gentler and more distant, and where its diverse sounds weave a quiet and personal harmony of their own.

**Shobhana Jeyasingh** has directed the Shobana Jeyasingh Dance Company since 1988. Her choreography for the Company includes *Raid* (1995), *Romance... with footnotes* (1993), *Making of Maps* (1992), *New Cities Ancient Lands* (1991), *Correspondences* (1990), *Defilé (1989) and Configurations* (1988).

Television includes *Duets with Automobiles* (Arts Council/BBC series), *Map of Dreams* (After Image Productions) and Dancing by *Numbers* (Bandung File) for Channel 4, *Late* for The Late Show and a Network East Dance Special for the BBC. Her work for theatre includes *Cyrano* and *The Little Clay Cart* for the Royal National Theatre. She has choreographed new work in Canada, Amsterdam and Finland.

Shobana and the Company have received numerous awards including Digital Dance Awards and the Prudential Award for the Arts in 1993. Shobana was awarded an MBE in this year's New Year's honours for services to dance. She also holds an honorary MA from Surrey University and an honorary doctorate from De Montfort University, Leicester.

# The South Bank Dance Papers

The 'Dance Paper' produced here is the result of a collaboration between the South Bank Centre and the dance department of Middlesex University. The South Bank Centre produces a year-round strand of talks, debates and seminars on dance and dance-related issues as an integral part of its dance programme and as such involve important dance practitioners and historians from Britain and abroad. Until now, the material generated in these events has had no life beyond that of the event itself. These 'Dance Papers' are the first time that elements of this material have been used to bring the issues to a wider audience and to preserve them for future audiences and dance students. The South Bank Centre and the Dance Department at Middlesex University share the view that the material presented in the context of a performance programme deserves a more substantial platform: these 'Dance Papers' provide this while deepening the debate and extending further the issues raised.

Shan Maclennan and Vicky Thompson
Southbank

*Performing Arts International*
1996, Vol. 1, Part 1, pp. 81–86
Reprints available directly from the publisher
Photocopying permitted by license only



# Depression in the Land of Dance Administrators

Julia Carruthers

What do dance administrators moan about when they get together? Why is the mood generally so despondent and weary? A brief look at self-esteem, the response from promoters, the use of meagre resources, and the lack of recognition by one who has been crazy enough to rejoin the club.

KEYWORDS: Dance administration, administrators morale, Arts Council funding, English and American dance administration.

Suzanne Walker, the administrator of V-TOL Dance Company and I are sitting in a crowded pub garden in Clapham. It is a beautiful, hot evening and there is a magnificent sunset starting. We are exchanging heart-warmingly disloyal and bitchy stories about the land of dance and dancers and we are laughing a lot.

"Well, you seem all right," I say, "Here I am trying to collect evidence to support the much repeated maxim that all dance administrators are depressed and you're sitting there looking tanned and healthy with a big grin on your face... I thought dance administrators were all supposed to have anxiously hunched shoulders, mumble wearily into their complimentary bottles of Beck's beer and constantly shake their heads."

At a recent meeting of the Management Liaison Group, a regular talkshop for administrators of small and middle-scale dance companies, Jane Attenborough (Director of Dance UK, the UK's umbrella organisation for Dance) offered to invite a psychologist to the next session, as she was worried about the dreadfully despondent mood amongst the group. She was not joking.

I ask Suzanne what she thinks some of the problems are. A deep sigh and slumping of the torso follows (you understand that dance people follow body language closely). She pulls a rumpled notebook out of her bag and flicks efficiently to where she has scribbled her findings. She has consulted with other dance administrator colleagues, having been warned by me that this was an item on the agenda for tonight's discussion, never mind the chilled white wine and pistachio nuts...

The start of the problem seems to be the low status of dance and its poor self-esteem. Hardly anybody in the real world (outside the arts administration mafia and a few stalwart journalists) has heard of the artists you are working with, never mind seen them. Rarely is what you are doing thought to be exciting or, God forbid, as important as working in television. People raise their eyebrows in consternation at the words contemporary dance. Your old school friends notice that when you go out to eat together you still choose the cheapest thing on the menu (pretending to be vegetarian) because you are probably earning half the amount they are. There are terrifying conversations over dinner (even in this post-Thatcher world) about how, really, one is expected to be earning one's age (i.e., if you are 30, you should be earning £ 30,000!). This all rather demoralising, even if it makes you feel like you have integrity.

The chief agony, chanted as a mantra by many working with the smaller and younger companies, is that when it comes to attempting to book performances for your company, those promoters and venue managers *never, ever, ever* ring back (Suzanne and I giggle about them maybe having phones that only accept incoming calls... but this is not possible, surely?). All dance administrators would confirm that these people need colossal amounts of chasing: you send packs, letters, faxes and you leave a thousand messages; they continue to be elusive and fail to respond. Gosh, it is hard to get those performances booked (never mind in some kind of coherent order in terms of geography and chronology). All the energy has to come from you and guess which cards are the most thumbed ones in the index box? The pressure is on, especially as all funding bodies continue to play the numbers game – prescribed numbers of dates for particular amounts of subsidy. The *only* time a promoter will contact you, quite exceptionally, is when they suddenly have a cancellation and an emergency on their hands.

Another serious concern amongst the administrative clan in dance is the lack of resources. The struggle with small budgets is constant. The issues here are many, but familiar themes will be the pathetic sums available for marketing, for commissioning other artists, for paying for additional help in the office and for much-needed taxis. The trepidation over the spending of every £50 is a terrible thing. If any budget projection does go over, you can be sure it won't be one of the administration items. Administrators are usually so frugal that savings are often made, allowing the production budgets to spill beyond their means...

The support systems for administrators are not as good as they could (or will) be. For example, many of the dance companies that have grown rapidly over the last 6 years either do not have Boards of Directors or have small, embryonic Boards. Boards tend to meet irregularly and do not really understand the issues. The meetings are pleasant and the friends of the artistic director are all clever and interesting people, but little is achieved and virtually nobody understands what is going on with the budget. Nobody has sufficient private wealth to write a personal cheque. Board members feel apprehensive about quizzing an exhausted administrator who is to be tabling yet another revised budget. And sometimes the artistic directors are abroad at the time of the Board meeting. The hunt for new, viable Board members and Chairs of Boards with suitable influence is constant.

It is taken for granted that the artists should gain status and recognition. Conversely, the administrators are rarely seen as being creative and powerful. An administrator's kudos might well be temporary, only bestowed upon them as long as they are associated with a particular dance company that is flavour of the month. The fact that the administrators are responsible for finding the money to realise the projects and, in several cases, being the brainchild behind it all, goes unremarked. Without the administrators securing that Arts Council or Regional Arts Board grant there would be no jobs for the company next year. One administrator confessed:

"Forgive me for sounding a weird combination of jealous, bored and bitter, but I am always being encouraged by the funding bodies to go and visit venues in order to chat up promoters personally, so they will give us a tour booking. I tell you, it's not *me* the venue manager is going to suck up to and fawn around but my artistic director whose aura is doubtless so much more exciting than mine."

Administrators will readily admit that they are not fighting for themselves or for their own glory, nor do they see themselves as charismatic showstoppers. In many cases it is accepted that, for example in the choreographer led companies, without the Artistic Director making work the company would no longer exist. The administrator is replaceable. Their job can be advertised and someone else found. It was also pointed out to me that the artistic directors can go off and earn money from pop promos and television, or other commissions while the martyrdom of the administrator is to stay plugging away in the office.

Anyone who came to observe one of Dance UK's Management Liaison Group meetings would immediately observe that there is an unhealthy gender balance, a monstrous regiment. Although the two 'national' ballet companies are run by men, practically every other dance company in the UK is organised by a woman. Dance is meant to be 'the female art-form' but it should cause disquiet that the dance situation so often resembles an old-fashioned home, with the woman behind the scenes, nurturing and encouraging, equipping and being practical. It all falls into line with the supposedly quaint idea that women are not used to putting themselves first. The irony is that the dance administrators may have to give up on the job once they have children, as the cost of childcare bears no rational relationship to the wages they are earning. This is just one of the worries about career development.

In the land that really gave birth to contemporary dance, America, administrators get called 'Executive Director'. They 'talk up' the 'product' they represent and its importance. In England the word 'administrator' implies that the people running the companies are secretaries. We English working in dance are modest and, comparatively, inarticulate. We're too shy and polite and we certainly haven't got all that American energy. A delegation of American dance administrators from companies such as Urban Bush Women, Bebe Miller and Doug Elkins visited the UK recently and met with a peer group of English dance administrators. The English crowd felt depressed and cowed by the dash and animation of the Americans. It wasn't actually as if the quality of the work was objectively superior or that the companies were doing better financially, but the Americans seemed to feel so much more healthy and optimistic about what they were doing (or they sure put on a damn good show and should be giving us acting lessons!). Of course, we must remember that in the United States someone working behind the counter at a car rental company will make the task of hiring you a car seem vital to the future of the human race and its sanity. This is because the whole psychology of America is that there is a pot of gold at the end of the rainbow. It appears to be too hard for the UK dance administrators to believe in that rainbow very often.

Those working elsewhere in dance or maybe on the Boards of dance companies and observing the administrators in action are liable make more general comments and possibly have an cooler-headed view:

"The trouble is that the administrator is left to cope with everything. They are always where the buck stops. Some of these dancers and choreographers are so helpless. The administrator is left keeping the Board, the dancers, the technical team, the composer, the sacked designer, the journalists, the photographer, the minibus driver and goodness knows who else happy. I've seen situations where the administrator has to cope with a smoke machine not working in a theatre 150 miles away from the office, because the artistic director and technician are incapable of taking constructive, speedy action. Or suddenly the lonely administrator is telephoned from Tokyo because the artistic director has overslept, missed his flight and wants his office to sort it out. No wonder these people get fed up and burnt out. Don't forget as well that the artists can enjoy the refreshment of funded "Research and Development" time, which the administrators don't get!"

I interviewed another administrator who I caught red-handed scanning the pages of The Guardian on a Monday (for the Creative and Media section, in other words the arts administration jobs). She was anxious to remain anonymous. Her summary of the issues was becoming familiar:

"The problem is that the same old problems keep coming up and it makes you weary. However advanced you get, you never stop having to do the mundane tasks. I consider myself pretty exprienced and quite a sophisticated operator, yet I still feel abominably patronised by many of the people I deal with. Maybe it's because I earn much less than they do and because I work in the artform that hasn't been allowed to grow up yet, called Dance. You are never able to keep up with the workload. It's like one of those dreams in which you are running and running to keep up and it's all hopeless. I often have fantasies about working for a Festival, because I don't mind working my guts out, but it would be nice to think it would all be over in three weeks! I used to want to work for a funding body to have a rest from it all and the pension, the expenses, the maternity leave were appealing – but it now seems far less attractive – all the funding officers I know keep complaining about filling in charts with targets and ticking boxes."

A comment from one of the funding officers was sympathetic:

"Although the opportunities for networking and meeting have improved tremendously, the smaller-scale dance administrators do all lead such isolated lives. They are rarely able to learn from anyone in a better position. Morale is buoyant enough on the large-scale and people still seem to have a sense of humour there, but the small and middle-scale workers have very little contact with the General Managers of the biggest companies. I think that the large-scale administrators are paid a lot more, but actually don't have that much more responsibility, they aren't carrying that much more weight. The worries are essentially the same."

Many of the dance administrators arrive in the jobs with little previous experience. They are young and hungry and are perhaps willing to earn a lower salary. There are few assistant administrator jobs in the dance world in which people can learn the ropes safely. There are many baptisms of fire. The fast track can be exhilarating, but suddenly the tarmac can run out and the road becomes awfully bumpy. People tend to stay in the jobs too long, as there are no obvious next steps. They are more likely to feel stuck nowadays, as companies rarely develop resources quickly and quantum leaps in public subsidy are virtually impossible.

Back to Suzanne and I in that bar in Clapham. The sun has now set. It is getting cold. We decide that we are working in dance because choreographers and dancers are some of the nicest, most wonderful people we have ever met (and a darn sight more congenial to work with than theatre directors, actors and musicians!). We wonder what some solutions might be. A summarised list of our curtailed discussion (we have no sweaters to put on, we have not been out with a wollen article of clothing for weeks) might run as follows. More dance links with the Independent Theatre Council (which provides excellent support to small-scale touring companies), better pay (somehow!), sabbaticals, better links with the large scale companies, offices in the same buildings as rehearsal studios, more staff in the office, a huge generic marketing campaign for contemporary dance, more support form the BBC and Channel 4 and more mobility in the dance jobs market. Suzanne and I have plenty of ideas — only some of them need money — Suzanne's parting shot is that my article is bound to be full of dreadful generalisations and that I'd better not wash too much dirty linen in public...

*Performing Arts International*
1996, Vol. 1, Part 1, pp. 87–91
Reprints available directly from the publisher
Photocopying permitted by license only



# Courting Culture and Contemporary, (sub)urban, Social Dance for Youth

Ruth Trueman

An examination of social dance activity as a courting mechanism in suburban night clubs and in specialised partner dance nightclubs.

KEY WORDS: Social dance, night clubs, youth, courting, feminism.

I became interested in this area of study as a regular club-goer with an interest in feminist politics. I wanted to understand the relationship between the dancing that took place in clubs with the 'romantic' activity which went on. I felt that dancing was a more powerful influence on the nature of these clubs than had been previously recognised. The type of clubs I visited most frequently were suburban nightclubs where the emphasis was on courting rather than dancing, music, drug or other concerns of city centre nightclubs.

There are two problems with this area of study; terminology and written resources.

Using the term contemporary, urban, social dance for youth (or CUSDY) is an attempt to group certain sorts of social dance activity together. This area of study might be called 'club dancing' or 'disco dancing'. However 'disco' is a very specific type of social dance, from a specific era to a specific type of music and, although the term is used widely to describe a number of social dance events where people dance to 'popular music', it is outdated and inaccurate. 'Club dancing' can be confused with dancing professionally in a club as entertainment for the club goers and so is also confusing.

Whilst there are many publications about popular music and youth culture, there is very little written about dance. Also whilst youth culture has been discussed since the 1960s, the role of young women in youth culture is often neglected. The work of Angela McRobbie (1991), Sue Lees (1986) and Helen Thomas (1994) has gone some way redressing this imbalance.

For this discussion it is assumed that the club goers are heterosexual and I do not differentiate between the behaviours of different ethnic groups, although there is an implicit examination of a certain class of youth by the nature of the venues that were studied. Youth, in this study, means men and women between 18 and 25, officially, and unofficially as young as 15 or 16 years old.

## *The social dance event as a mating ground*

Social dance spaces from dancehalls to clubs have been considered to be places to meet new people or, more specifically, new partners. Nightclubs are described by the users as 'cattle markets', 'pick-up joints', 'pulling places' and so on and Mungham (1976) aptly describes them as places where 'youth can pursue itself'.

Tracing the historical development of social dance since 1945, it is clear to see the move away from partnered and set dances to individual and improvised dances – a trend which began in the 1960s as part of the ethos of the time. Although there were some partnered dances as part of the 'disco' era in the 1970s (for instance the Spanish Hussle), since the 1960s, few set/partner dances have been prevalent and none appear on the dance-floors in the early 1990s.

Since my original research, several Latin-American dance crazes have had an influence on urban dance-floors. Certain venues have a regular Salsa night where a two hour class is followed by an evening of Salsa dancing. I maintain, however, that people dancing in the mainstream suburban nightclubs do not dance with a partner.

## *How does dance work as a courting mechanism in suburban nightclubs and speciality partner dance venues?*

My findings showed very clearly that women danced a great deal more than men in suburban nightclubs and that women were considered better dancers than men. The statistics I collected show that two-thirds of nightclub attendees are men and that two-thirds of the people on the dance-floor are women. This immediately sets up the familiar dynamic of the nightclub where the majority of men watch the women dance.

There are many well rehearsed arguments about why men don't dance (dance as a 'feminine art form/activity because it is both expressive and body centred) which I will not examine here. I see indications of this image of dance changing. There is now a much greater interest in the male physique shown in advertising and in men's magazines, the partner dance crazes, male roles models dancing (in the Guinness advertisement and the credits to the television series *Men Behaving Badly)* and the explosion of all male bands who dance as an intrinsic part of their performance (Take That and so on).

Whilst questionnaire data refutes that the women dance for the men's approval, interview data and observation suggest otherwise. The interviewees stated that they made more effort to dance well if they knew someone, especially if it was a man, was watching them. There is a limit, however, to what you can do to attract a non-dancing potential partner.

All the available data suggests that women are much less likely to initiate contact with a non-dancing, potential partner than men. Once again there are well rehearsed reasons for this, to do with the continuing stigmas against women who show 'too much' interest in men – 'slags'. The women interviewed stated the various means by which they would attempt to attract attention from potential partners who were dancing, but the majority agreed that they would not use a direct, verbal approach to either dancing or non-dancing potential partners.

The two main strategies used to attract a man's attention whilst dancing are as follows:

1 Copying or responding to the other person's movements. This is used as part of a general indication of friendship and can be clearly seen among groups of friends dancing. It is good as an initial link. It shows observation, sensitivity and interest in the other person. What it will not indicate is any sexual interest as it is recognised as being a friendly exchange.

2 Dancing close to someone, accidentally touching someone or bumping into them.

This is clearly a more flirtatious strategy involving actual (or near) body contact. Interviewees stated that this would only occur after a period of copying/responding or after other indications that the male counterpart was interested. Observation shows that a crowded dance-floor gives quite a bit of opportunity for this courting strategy to be explored.

Although these two dancing strategies have been expressed by a number of interviewees, the majority of experiences indicate that it is not until eye contact is established that any indication of interest can be registered. In this way, the non-dancing potential partners are still able to partake in the courtship activity by watching a dancer that they like and waiting for her to 'catch their eye'. Repeated glances are an indication of interest and are used as a prelude to verbal contact which is usually in the form of asking the woman 'to dance' or to have a drink.

Given the individual nature of the dance activity, asking someone to dance is simply part of an out-dated ritual. What it means in practice is an agreement to dance opposite each other, increase the level of copying/responding and possibly some physical contact. Withdrawal of copying/responding when dancing is an indication of disinterest and a sensitive partner may back-off at this point. However, given the emphasis on being a 'nice girl' which is still prevalent, the male may just take this as coyness or shyness and will not take the hint.

The codes of behaviour used in partner dances to facilitate courting in a social dance environment, have been replaced by a complicated set of non-verbal signals that are open to mis-or non-interpretation. Social dancers recognise that a Salsa club where men and women dance in couples is a much better 'hunting ground' than a traditional nightclub. Typical responses from club goers interviewed recently include:

"You might get to dance with nine guys during an evening and you can choose which one – if any – you want to go home with them... I'll only go clubbing now if I want to dance with my friends."

Clare, aged 24, 1995

"If you want to get-off with someone – go to a Salsa club! Its a guaranteed pull."

Emma, aged 20, 1995

The reason for this is that the couples get direct, physical contact with each other and can collect a great deal of information to base a decision about their suitability as a partner. The information is collected visually, as it would be in an ordinary night-club verbally as there is a legitimate reason to communicate, physically as the couple move together in a taught and therefore mutually agreed pattern and experientially – how is it to be with this person?

I am interested in how my feminist interviewees feel about waiting to be asked to dance and about being led around the dance-floor. Some confess a liking for the formality of the etiquette and to the dance experience. Others have commented that this generation of men treat leading a women differently to their fathers as they are only just being taught to lead and are used to seeing women dancing on their own.

"Young men lead really badly, but it is OK because that way I know they don't presume always to lead – to lead in everything."

Elizabeth, aged 26, 1995

## *Conclusion*

We have a culture of diversity and choice and it seems appropriate that there be various types of CUSDY events for young people to choose from. The courting culture of nightclubs is still most noticeable in suburban areas where the choice of venue is diminished, where there may be no partner dance venue nearby. For these young people, the inadequacies of the dance-floor as a mating ground are mingled with the confusing messages gleaned about relationships and sex from *Baywatch* and *Blind-Date* versus *Crimewatch* (rape/attack reconstructions) and HIV warnings.

It will be interesting to see if the Salsa craze goes beyond the urban into the realms of the suburban social dance scene and, if it does, what impact that will have on CUSDY events as mating grounds.

## *Bibliography and References*

Lees, S *Losing Out: Sexuality and Adolescent Girls* London: Hutchinson, 1986.

McRobbie, A *Feminism and Youth Culture: From Jacky to Just Seventeen.* London: Macmillan, 1991.

Mungham, G 'Youth in Pursuit of Itself' in Mungham, G and Pearson, G (eds) *Working Class Youth Culture.* London: Routledge & Kegan Paul, 1976, pp. 82–104.

Thomas, H *Dance, Gender and Culture* London: Macmillan, 1994.

*Performing Arts International*
1996, Vol. 1, Part 1, pp. 93–102
Reprints available directly from the publisher
Photocopying permitted by license only



# Signing in the Monsoon

Nigel Warrack

In travelling to the Nambikkai Foundation, South India, to work with young deaf adults, it has been possible to note and reflect upon their position and, by contrast, the equivalent in England. As a teacher and practitioner, I hoped to futher ideals promulgated by G. Sankala Pillai. Although an unexpected change of site occurred, newly organised workshops provided valuable insights. Tracking progress by use of a personal diary, the steps to achieving a dance/theatre dialogue with the handicapped from another culture can be seen.

KEY WORDS: Community dance, community theatre, south India, disability, deafness.

The Nambikkai Foundation is a rural community project in South India. The agricultural land is used for growing coconuts and limes, while the main complex houses dairy, poultry and goat units. All this provides resources for the area around Kanyakumari, the "Land's End" of India, where the Bay of Bengal meets the Indian Ocean. But the main function of Nambikkai is to provide education, training and employment opportunities for young deaf adults.

I first heard of the project in 1991 while working at another training institution, Rangha Prabhath Theatre Complex, 87 kilometres North, in Kerala. Rangha Prabhath was founded by the late G. Sankala Pillai for "the awareness and propagation of Gandhian ideals and values through drama among children and youth, particularly among the rural masses and weaker sections." There seemed to be potential for a connection between the two institutions.

I had been profoundly impressed with the vision of G. Sankala Pillai, his belief that no community was complete without a theatre along with a water supply, cultivable land and a place of worship. The creative energy of children, who "love to learn, but hate to be taught" is essential to the performances and the sense of co-operation between children and highly trained actors produces magical theatre. Much of my own work in the UK has involved disabled people, particularly with integrated groups such as Amici and Green Candle, and my intention on this project with Nambikkai and Rangha Prabhath, hoping to open up the possibility of an ongoing programme of arts with disability in Kerala.

There are many reasons why disabled people should be involved in theatre, anywhere in the world. As well as the need for equality of opportunity, surely a social imperative, there are certain specific areas in which diasbled performers are uniquely placed to deepen the artistic experience. In the area of dance and creative movement, for example, people with learning difficulties often have the upper hand. Frequently they are less inhibited, more spontaneous – the moment between feeling an emotion and responding to it in movement is less cluttered with critical analysis. Deaf culture has also contributed a great deal to British theatre, not least a physical language complete with its own grammar, history and poetry. In this respect contemporary British theatre does seem to be more developed than its Indian counterpart, in marked contrast with other areas of the arts, notably the vibrant execellence of children's theatre in Kerala.

Though it still has some way to go, the disability movement has made some significant inroads to British society. Things are also changing in India, but some of the issues are different. Many British disabled people say that the only things that really hold them back are other people's attitudes. Disability, often through poverty, is more common and certainly more visible in India, and as such it is accepted as a fact of life without the hushed tones and embarrassed glances that many western disabled people have to endure. But fewer disabled people are in full employment in India. Geraldine Dunlop, a deaf Englishwoman who works at Nambikkai, says that deaf adults in India have the same status as deaf children in the UK.

Kerala, however, is probably the most enlightened Indian state: the first place in the world freely to elect a communist government (in 1957), it also has the most active women's movement and a literacy rate of 91%. The caste system, according to Cuckoo Parameswaran, an actress who trained with G. Sankala Pillai, is only really observed by the oldest generation. Certain aspects remain, however: "If a Brahmin who has disability paints a picture, he is immediately hailed as Picasso." There is no directly comparable situation in the UK, although I was reminded of a remark by Chantelle Nassari, a dancer with Green Candle. Even though she has had many favourable press notices for her work as a professional performer, the copy heading always reads something like "Wheelchair Girl Makes Good". In both societies it seems that disabled people are individually labelled either as heroes or as victims, but never have the freedom to be unremarkable.

*There follows a personal diary of the two week project.*
Sunday 13 August: First meeting at Rangha Prabhath. Kochunaryana Pillai, the director greeted us warmly as the children presented us with garlands of flowers, fresh coconut juice and red bananas. Kochunaryana Pillai speaks very little English, I still less Malayalam, so the meeting was translated by the recently appointed administrative consultant. It quickly emerged that Rangha Prabhath had changed. I was asked to pay a fee to each of the 'child artistes' attending my workshop. We were then informed that the collaboration with Nambikkai was not in line with the programme of work at Rangha Prabhath. G. Sankala Pillai's original vision of "Gandhian ideals and values through drama among children and youth particularly among the rural masses and weaker sections" seemed to have been overlooked. His work is being treated as a marketable product at the expense of long term growth. However, artistic standards are extremely high in Keralite children's theatre and the result is an effective boycott of Rangha Prabhath by ex-students.

Monday 14: Fortunately for me two ex-students, Babu Kuruvilla and Cuckoo Parameswaran were in the process of setting up a new project, a programme of workshops in the coastal area of Anjengo, in a day centre for so-called "drop-out" children, who weren't at school because of problems at home. This sounded like a more fruitful collaboration. We arranged to visit on Wednesday.

Wednesday 16: Anjengo is a tiny fishing village near the Allepey backwaters, half way up Kerala. The "drop-out" centre is run by

Learning the Welly Boot Dance.

Christians (St. Thomas the Apostle is said to have landed at the port of Miziris in 54 AD and now nearly a quarter of Kerala is Christian) and there is a daily Mass, although many of the children came in with a red dot on the forehead from morning 'puja' in the local Hindu temple. But they seemed happy to be there, and the nuns were extremely helpful.

Thursday, Friday, Saturday: We played for three days. The children had had no experience of theatre and gave the simplest exercises a delightful freshness and enthusiasm. They particularly liked Scottish Country Dance. Dancing in couples was a completely new concept as was the use of space, but some of the actual steps related to their own folk dances (based on Mudi Yattu, a major influence on Kathakali, the classical dance-drama). As all South Indian dance is performed with bent knees and flat feet, the Scottish straight legs and pointed feet were adapted to produce a charming new dance.

Monday 21: By car to Nambikkai with Cuckoo and Babu. Lengthy discussion on what form the final performance should take. One argument was for an open workshop. Babu in particular was very keen to aim for a clearer narrative structure. With so many unknowns – the Anjengo kids being away from home for the first time, the deaf students' first experience of theatre – we agreed to take things as they came.

On arrival we were fed in the canteen (boys and girls separate). The students use a mixture of Tamil and British Sign Language (BSL). One thing that is certainly not a problem in Nambikkai is political correctness. The most common sign for India is rubbing a finger on the wrist to indicate skin colour. As this didn't work on my white skin, I tried a version I had been taught before: two index fingers pointing downwards to form a V to indicate the shape of the country. I caused great hilarity by accidentally joining my thumbs at the top, apparently the sign for vagina. We also discovered that the sign for Thiruvananthapuram, the capital of Kerala is the same as the sign for crazy, while "Madras" and "toilet" are interchangeable. We were shown our sleeping quarters (also separate, even for married couples) and introduced to the directors. We were repeatedly cautioned on the dangers of physical contact among the students, based on "bitter experience" in the past. I've come across this many times in the UK – care staff who have a horror of disabled people touching each other – but it seemed strange coming from an organisation which is essentially about disabled rights, not charity.

Tuesday 22: Anjengo kids arrive. Before meeting the Nambikkai students we were presented with a short list of 15 who had been specially chosen. I suggested that the workshop should be open to anyone who was interested. Big mistake. The result was group of 43 deaf teenagers with 6 astonished hearing children, as my instructions were simultaneously

Group warm-up at Nambikkai.

Group warm-up at Nambikkai.

translated into Malayalam, Tamil and BSL. Most of the games took longer to explain than to play. Only two things worked: a Chinese-style warm-up (borrowed from an early morning keep-fit session in Tianenmen Square) and a version of the South African Zulu gum boot dance.

Nigel Warrack losing a game of Puchaillikali.

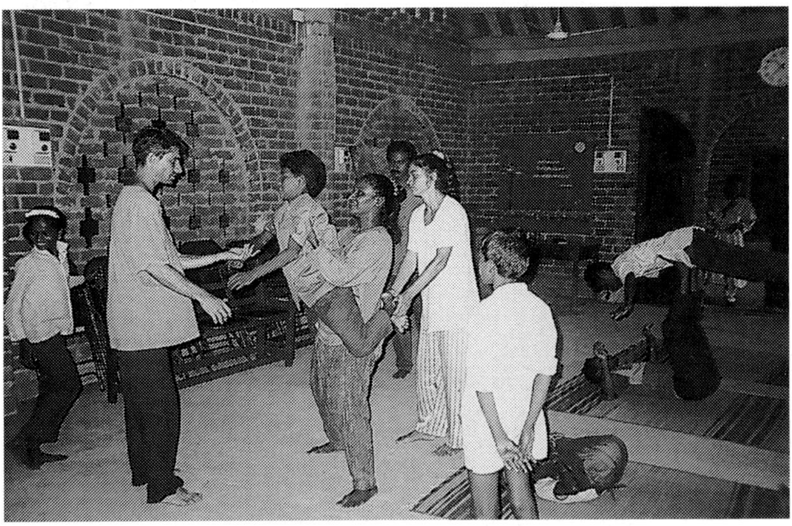

Acrobatic balance at Nambikkai.

Wednesday 23: The Anjengo kids seemed tired and rather unnerved by yesterday's experience. Pradheep, their nanny, asked if they could just watch today. I began to panic; was I asking too much to them? We worked with the original short-listed 15 on very basic theatre games:

"The Wind of Death".

Nigel Warrack with students at Nambikkai.

mirrors, trust exercises etc. People seemed to be enjoying themselves. The really pleasing moment was when Dolphin, the oldest of the Anjengo kids, asked if they could please join in. The prospect of having fun overshadowed any previous worries about disability or age difference. We finished the day with a version of Bob Marley's "Don't Worry" in English, Malayalam and BSL.

Thursday 24: Much better. Starting with some traditional Kerala games – Puchaellikali (a sort of 'cat and mouse') and Narnga Pal (similar to 'oranges and lemons' but finishing in a tug of war between women and money) – we went through a western contemporary dance warm-up, simple acrobatic balances, some Kathakali facial expressions and gestures, then a long structured improvisation based on movements and images from the students' home town or village. There was absolutely no problem with boys and girls working together. No sign of 'rascalish behaviour' that we had been warned about and certainly no 'timidity' from the girls, some of whom were getting rather bossy. The flirting that was going on clearly had less to do with disability and everything to do with adolescence.

Friday 25: Bit of a disaster. Spent the morning trying to shape the group improvisations into a narrative structure. Somehow as the groups merged, details got lost and people got bored, then tired. By lunchtime we had reduced yesterdays's enthusiasm to an embarrassing mess of young people pretending to be trees. Recovered a bit in the afternoon with some of the material that had worked yesterday.

Saturday 26: Went back to the idea of an open workshop. Much better atmosphere, with some excitement at the prospect to showing to the rest of the centre. When it came to it, showing was a huge sucess. The audience enjoyed the improvisations but by far the most successful part was the open warm-up; the students seemed really interested to know

Rehearsal at Nambikkai.

what their colleagues had been doing all week. I have never been so clearly shown the value of process over product in community dance. The embarrassing farewell speeches were, luckily, cut short by a brief reappearance of the South West monsoon.

It will, by this time, be apparent that some aspects of this fortnight were substantially different from the project I had planned.

Firstly, the unstated policy shift at Rangha Prabhath had made it seem unproductive to force a collaboration with Nambikkai, which in a sense had been the whole point of the trip. However, the refusal of G. Sankala Pillai's ex-students to work there, as well as being a testament to the depth of this vision, has had some positive results. Rangha Prabhath is an outdoor theatre complex built around the mango tree under which G. Sankala Pillai conducted the first work shops for the children of the local village. (The tree still stands, upstage

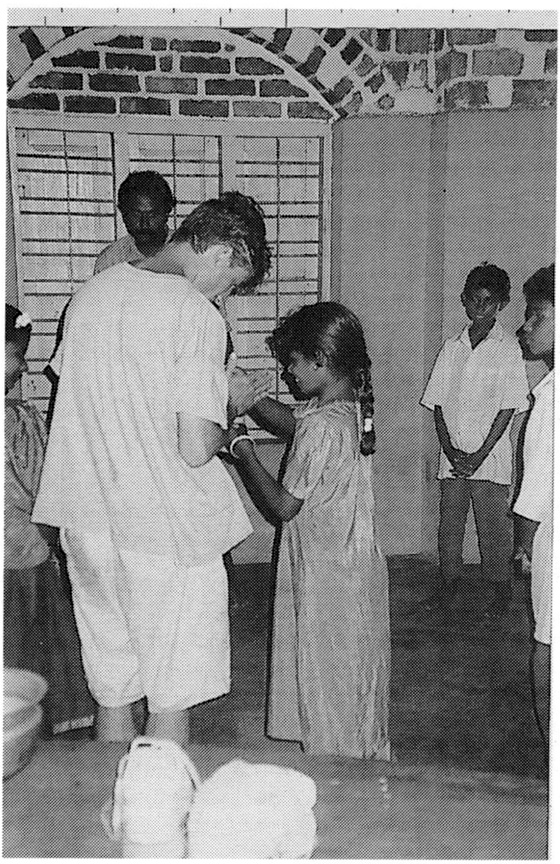

Nigel Warrack with Sherri from Anjenga.

left). The reality is that the workshop programme in the coastal area of Anjengo is actually closer to Pillai's original intention than the high profile "presentations", exhibitions and seminars/discussions" led by Rangha Prabhath's new administrative consultant. Closer, too, are other projects set up by ex-students such as one in Koothuparamba, Northern Kerala, with Hindu and Muslim children and another in the so-called "tribal area" near Mannaghat in the inland hills. Secondly, the children who eventually travelled from Anjengo to Nambikkai with us were much younger than the ones who would have come from Rangha Prabhath. Aged between seven and nine, never having left their village, they clearly had different experiences from deaf teeenagers in an institution. What they had in common, however was imagination, enthusiasm and a total lack of preconceived ideas about theatre. It was an inspiring group to work with, as Cuckoo and Babu continue to do.

Thus, in spite of all these changes in plan, my original hope – to initiate an ongoing programme of arts with disability in Kerala – seems likely to be realised. Anyone with practical experience in community dance will be used to reworking or even aborting plans at a moment's notice. This, for me, has always been one of the most exciting aspects of the work: the fact of suddenly being forced to communicate in a spontaneous direct and real way is a useful lesson to an artist in any discipline. Merce Cunningham describes dancing as the art of the present tense. "It bridges a point in human behaviour somewhere between the intellect and the instinct. This is why it fascinates people." This brilliant description of the essence of an art form is as relevant to a small child in India as it is to an international genius such as Cunningham.

If theatrical activity is to have any depth, it must involve co-operation as well as techinique and creativity. Community dance requires enormous co-operation from all involved and the results can be moving, spectacular, even touchingly comic. In this respect it is essentially an artistic experience. The social benefits, which are undeniably important, are a result rather than a cause. This concept is well established in South India, with the result that children's theatre has a status comparable to the mainstream, sometime that British funders and promoters could and should learn from. On the other hand, the idea of art with and for disabled people – one of the really exciting growth areas in British theatre of the last decade – is virtually unexplored in South India. The possibilities for artistic exchange continue to grow.

*Performing Arts International*
1996, Vol. 1, Part 1, pp. 103–104
Reprints available directly from the publisher
Photocopying permitted by license only



# Notes on Contributors

**David Hughes** has taken up his appointment as Senior Lecturer at the Department of Visual and Performing Arts, Nottingham Trent University. His paper developed over time through conferences (South Bank in particular), and focussed through further research with colleagues.

**Alexandra Carter** is principal lecturer at Middlesex University's School of Dance. Her PhD combined interests in new historiography, feminist methodology, and dance analysis. Since then, her research has developed in the application of feminist perspectives to dance analysis and history. One current interest is in finding new ways of teaching dance history.

**Emmerich Simoncsics, Evelin Gruber, Andrea Kocevar, Renate Hammer, Michael Fiala, Peter Parolek** formed the Applied Aesthetics research project at the Technical University of Vienna. Univ. Doz. Dr. Emmerich Simoncsics is chairperson of the project. It is an interdisciplinary project between architects, musicians, sculptors and dancers. Their aim is to find connections in the creating principles of these disciplines.

**Michael Bridger** is Professor of Music and Head of School at Middlesex University. His main research is the study of electro-accoustic music, within broader fields of interest in postmodernism and critical theory related to technologically based visual and performing arts. He also has a long standing involvement in music pedagogy, both at the level of professional training of school music teachers and in methodological research.

**Toshihiko Norizuki** is based at Tamagawa University in Tokyo, in the Drama office of the Arts and Education Department.

**Shobana Jeyasingh** is an internationally-known choreographer working in the field for many years. Her company operates from The Place Theatre in London.

**Julia Carruthers** has extensive arts administration experience and currently is an officer at the Arts Council of England's Dance Department.

**Ruth Trueman** is Lecturer in Dance at Middlesex University, her main research area is in contemporary social dance (for youth). Other interests are in gender, community dance, and arts funding. Until recently Ruth was also a Dance Officer at the Arts Council of England.

**Nigel Warrack** is Lecturer in Dance at Middlesex University, with interests in community dance, dance and disability, and contemporary dance.

*Performing Arts International*
1996, Vol. 1, Part 1, pp. 105
Reprints available directly from the publisher
Photocopying permitted by license only



# Index

*PERFORMING ARTS INTERNATIONAL*
*AN INTERNATIONAL JOURNAL*

# Notes for contributors

Submission of a paper will be taken to imply that it represents original work not previously published, that it is not being considered for publication elsewhere and that, if accepted for publication, it will not be published elsewhere in the same form, in any language, without the consent of editor and publisher. It is a condition of acceptance by the editor of a typescript for publication that the publisher automatically acquires the copyright of the typescript throughout the world. It will also be assumed that the author has obtained all necessary permissions to include in the paper items such as quotations, musical examples, figures, tables etc. Permissions should be paid for prior to submission.

**Typescripts.** Papers should be submitted in triplicate to the Editors, *Performing Arts International c/o* Harwood Academic Publishers, at:

| | | | | |
|---|---|---|---|---|
| 5th Floor, Reading Bridge House<br>Reading Bridge Approach<br>Reading RG1 8PP<br>UK | or | PO Box 27542<br>Newark,<br>NJ 07101-8742, USA | or | 3-14-9, Okubo<br>Shinjuku-ku<br>Tokyo 169<br>Japan |

Papers should be typed or word processed with double spacing on one side of good quality ISO A4 (212× 297 mm) paper with a 3 cm left-hand margin. Papers are accepted only in English.

**Abstracts and Keywords.** Each paper requires an abstract of 100–150 words summarizing the significant coverage and findings, presented on a separate sheet of paper. Abstracts should be followed by up to six key words or phrases which, between them, should indicate the subject matter of the paper. These will be used for indexing and data retrieval purposes.

**Figures.** All figures (photographs, scheme, charts, diagrams and graphs) should be numbered with consecutive arabic numerals, have descriptive captions and be mentioned in the text. Figures should be kept separate from the text but an approximate position for each should be indicated in the margin of the typescript. It is the author's responsibility to obtain permission for any reproduction from other sources.

*Preparation:* Line drawings must be of a high enough standard for direct reproduction; photocopies are not acceptable. They should be prepared in black (India) ink on white art paper, card or tracing paper, with all the lettering and symbols included. Computer-generated graphics of a similar high quality are also acceptable, as are good sharp photoprints ("glossies"). Computer print-outs must be completely legible. Photographs intended for halftone reproduction must be good glossy original prints of maximum contrast. Redrawing or retouching of unusable figures will be charged to authors.

*Size*: Figures should be planned so that they reduce to 12 cm column width. The preferred width of line drawings is 24 cm, with capital lettering 4 mm high, for reduction by one-half. Photographs for halftone reproduction should be approximately twice the desired finished size.

*Captions*: A list of figure captions, with the relevant figure numbers, should be typed on a separate sheet of paper and included with the typescript.

*Musical examples*: Musical examples should be designated as "Figure 1" etc., and the recommendations above for preparation and sizing should be followed. Examples must be well prepared and of a high standard for reproduction, as they will not be redrawn or retouched by the printer.

In the case of large scores, musical examples will have to be reduced in size and so some clarity will be lost. This should be borne in mind especially with orchestral scores.

**Notes** are indicated by superior arabic numerals without parentheses. The text of the notes should be collected at the end of the paper.

**References** are indicated in the text by date name and the system either "Recent work (Smith & Jones, 1987, Robinson, 1985, 1987) ..." or "Recently Smith & Jones (1987) ..." If a publication has more than three authors, list all names on the first occurence; on subsequent occurrences use the first author's name plus "et al." Use an ampersand rather than "and" between the last two authors. If there is more than one publication by the same author(s) in the same year, distinguish by adding a,b,c etc. to both the text citation and the list of references (e.g. "Smith, 1986a") References should be collected and typed in alphabetical order after the Notes and Acknowledgements sections (if these exist).
Examples:

Benedetti, J. (1988) *Stanislavski*, London: Methuen

Granveille-Barker, H. (1934) Shakespeare·s dramatic art. In *A Companion to Shakespeare Studies*, edited by H. Granville-Barker and G. B. Harrison, p. 84. Cambridge: Cambridge University Press

Johnston, D. (1970) Policy in theatre. *Hibernia*, **16**, 16

**Proofs.** Authors will receive pageproofs (including figures) by air mail for correction and these must be returned as instructed within 48 hours of receipt. Please ensure that a full postal address is given on the first page of the typescript so that proofs are not delayed in the post. Authors' alterations, other than those of a typographical nature, in excess of 10% of the original composition cost, will be charged to authors.

**Page Charges.** There are no page charges to individuals or institutions.